Homology Molecular Modeling - Perspectives and Applications

Edited by Rafael Trindade Maia, Rômulo Maciel de Moraes Filho and Magnólia Campos

Published in London, United Kingdom

IntechOpen

Supporting open minds since 2005

Homology Molecular Modeling - Perspectives and Applications
http://dx.doi.org/10.5772/intechopen.91624
Edited by Rafael Trindade Maia, Rômulo Maciel de Moraes Filho and Magnólia Campos

Contributors
Vladena Bauerová-Hlinková, Jacob A. Bauer, Budhayash Gautam, Anupam Nath Jha, Sanchaita Rajkhowa, Zaved Hazarika, Gugan Kothandan, Ananthasri Sailapathi, Seshan Gunalan, Kanagasabai Somarathinam, Diwakar Kumar, Margarita Miranda-Hernández, Abraham Vidal-Limon, Guillermo Antonio Huerta-Miranda, Wendy I. García-García, Vishwambhar Vishnu Bhandare, Rafael Trindade Trindade Maia, Magnólia De Araújo de Araújo Campos, Rômulo Maciel de Moraes Filho

First published in London, United Kingdom, 2021 by IntechOpen
IntechOpen is the global imprint of INTECHOPEN LIMITED, registered in England and Wales, registration number: 11086078, 5 Princes Gate Court, London, SW7 2QJ, United Kingdom
Printed in Croatia

British Library Cataloguing-in-Publication Data
A catalogue record for this book is available from the British Library

Additional hard and PDF copies can be obtained from orders@intechopen.com

Homology Molecular Modeling - Perspectives and Applications
Edited by Rafael Trindade Maia, Rômulo Maciel de Moraes Filho and Magnólia Campos
p. cm.
Print ISBN 978-1-83962-805-4
Online ISBN 978-1-83962-806-1
eBook (PDF) ISBN 978-1-83962-811-5

For EU product safety concerns:
IN TECH d.o.o., Prolaz Marije Krucifikse Kozulić 3, 51000 Rijeka, Croatia, info@intechopen.com or visit our website at intechopen.com.

Meet the editors

Dr. Rafael Trindade Maia studied biological sciences at the Federal Rural University of Pernambuco, Brazil (2005), received a Master's degree in genetics, conservation, and evolutionary biology from the National Institute of Amazonian Research, Brazil (2008), and a Ph.D. in animal biology from the Federal University of Pernambuco, Brazil (2013). He is currently an adjunct professor at the Center for the Sustainable Development of the Semi-Arid (CDSA) at Federal University of Campina Grande (UFCG), Brazil. He has experience with population genetics, bioinformatics, molecular docking, homology modeling, and molecular dynamics of proteins. He works in the area of science and biology education. He leads the research groups Computational and Theoretical Biology (CTB) and Education in Sciences and Biology (ESB).

Dr. Rômulo Maciel de Moraes Filho holds a BA in biological sciences from the Federal University of Pernambuco, a Master's in agronomy with a focus on plant breeding from the Federal Rural University of Pernambuco and a Ph.D. in genetics from the Ribeirão Preto Medical School-USP. He has experience in population genetics, conservation genetics, conservation of genetic resources, bioinformatics and cultivation of plant tissues. He is currently a postdoctoral fellow at the Federal Rural University of Pernambuco, conducting research aimed at molecular characterization and selection of genotypes of the genus Psidium spp. resistant to nematodes.

Magnólia de Araújo Campos is a biologist, has a Master's degree in agronomy/plant breeding from the Federal University of Pelotas, Brazil, and a Ph.D. degree in biological sciences/molecular biology from the University of Brasília (2002). She has five years of experience in genomic sciences as a postdoctoral researcher at the Federal University of Lavras/Agronomic Institute (IAC), Brazil. She has been a professor at the Federal University of Campina Grande (UFCG), Brazil since 2008. She mainly works on the following topics: genomics, bioinformatics, tissue culture and plant cells, genetic transformation of plants, gene expression during plant-microbe interactions, and expression of heterologous proteins in bacteria.

Contents

Preface

This book is a select collection of works on the homology modeling technique, also called comparative modeling.

Molecular modeling by homology is a technique that has been growing and becoming popular in several fields of science such as structural bioinformatics, theoretical and biochemical chemistry, and computational biophysics.

Through this tool, it is possible to predict the three-dimensional structure of a protein from its primary sequence of amino acids, as long as there is a three-dimensional structure already known of a homologous protein to be used as a template.

The theoretical models obtained by the homology modeling technique have numerous uses. One is that knowledge of the 3D structure of a protein is crucial for understanding its function and behavior. Thus, theoretical models of proteins are formidable for a good understanding of biological systems and intra- and extracellular processes of living organisms.

Other applications are in the field of biotechnology and drug discovery. Theoretical models of target proteins for drugs or with applicability in genetic engineering can accelerate several discoveries of importance for biotechnological and therapeutic advances. Simulation studies of docking and molecule dynamics using theoretical models have grown enormously.

This book is aimed at researchers and academic students in areas related to molecular modeling, biotechnology, and molecular biology. It consists of seven chapters carefully selected and reviewed.

Section 1 contains the introductory chapter (Chapter 1), in which the editor and co-editors of the book make a synthetic and objective approach to the technique of comparative modeling, presenting this tool to the reader.

Section 2 presents two excellent review chapters that address fascinating topics in molecular modeling. Chapter 2 presents the Normal Mode Analysis technique, with examples and excellent didactic-scientific description. Chapter 3 takes a great approach to using force fields to validate three-dimensional models.

In Section 3, readers will be able to find scientific works with in silico experiments that used the modeling technique. This section consists of two chapters, where readers can find practical use in real academic work using molecular protein modeling techniques.

Section 4 contains two chapters that address applications associated with modeling.

With these works, readers will have access to information at the forefront of molecular modeling, making it possible to expand their knowledge and view on biological and biotechnological research.

Rafael Trindade Maia
Federal University of Campina Grande,
Sumé-Paraiba State, Brazil

Magnólia de Araújo Campos
Federal University of Campina Grande,
Cuité-Paraiba State, Brazil

Rômulo Maciel de Moraes Filho
Federal Rural University of Pernambuco,
Recife-Pernambuco State, Brazil

Section 1

Introduction

Chapter 1

Introductory Chapter: Homology Modeling

Rafael Trindade Maia, Magnólia de Araújo Campos and Rômulo Maciel de Moraes Filho

1. Introduction

Proteins are macromolecules present in all living beings and perform a huge variety of complex and diverse functions and structures. They are polymers of amino acids synthesized in the cell of living organisms, also called polypeptides. Determining the three-dimensional structure of a protein is crucial for understanding its function. However, experimental techniques for structural elucidation such as X-ray critalography and nuclear magnetic resonance (NMR) are complicated and expensive [1]. In this context, computational techniques for building structural models are a very useful and viable alternative for different situations. Among computational techniques, homology modeling, also known as comparative modeling, is the most used *in silico* tool for obtaining structural protein models, achieving excellent results [2].

Proteins are organized at different levels of structural complexity: 1) primary structure; 2) secondary structure; 3) tertiary structure; 4) quarternary structure (**Figure 1**). The primary structure of a protein comprises the linear sequence of the amino acids that compose it, with one end containing the carboxyl group of the first amino acid in the chain (C-terminal) and with one end containing the amino group of the last amino acid in the chain (N -terminal). The primary structure of a protein can be represented by a pattern of letters that represents its peptide constitution (amino acids). The secondary structure of a protein is determined by the primary sequence, which is decisive in the arrangement of the monomers (aminoacids) with each other and with the solvent, forming standard structures in three groups: the turns, the helix and the β-leaves. The way in which these secondary structures are organized three-dimensionally in space is what is called a tertiary structure, which is associated with the biological function of the molecule in question. In multimeric protein complexes (dimers, trimers, tetramers, etc.) there is also the formation of the quarternary structure, which is the oligomeric state formed by the aggregation of these macromolecular compounds of tertiary structure.

There are three types of computational modeling for predicting protein structures: by *ab initio*/*De novo*, by *Threading* and by homology modeling. Homology modeling is based on the premise that the three-dimensional structure of a protein tends to be much more conserved than its primary structure. Therefore, changes in the sequence do not always change the structural domains of a protein, thus maintaining its original function. It is assumed that proteins from the same functional family maintain their structural domains, which allows the so-called comparative modeling (by homology). If two proteins are homologous, it means that they belong to the same genetic and functional family, and hypothetically, they have the same structural motifs. In the case of a specific protein that does not have an elucidated three-dimensional structure, but it is homologous to a protein with a

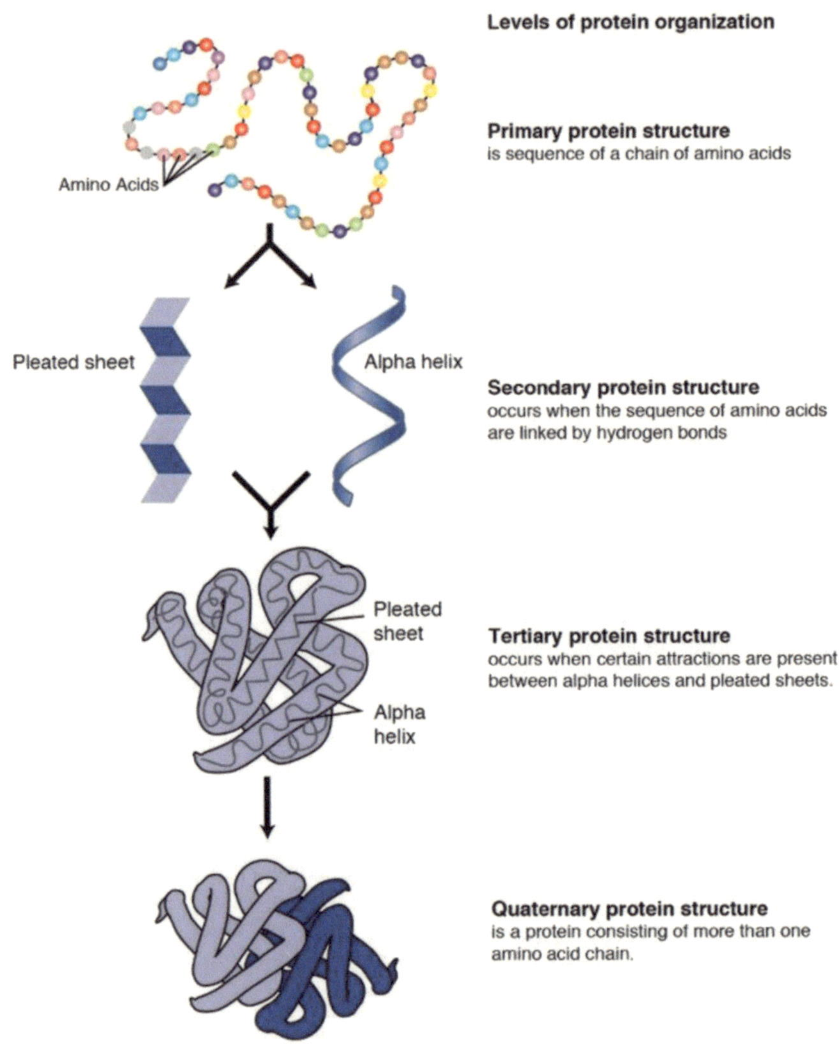

Figure 1.
Illustrative scheme for the structural complexity levels of proteins. Source: Google images.

solved structure, a three-dimensional model for the sequence can be built using the known structure as a template. As a rule, a minimum identity of 25% between the amino acids of two proteins is sufficient for the construction of models by homology. Sequence identities above generally 40% provides good models, while those above 50% tend to provide excellent theoretical structures [3].

However, in addition to the identity and similarity between the amino acids, other parameters must be observed when choosing a good template, such as the resolution in angstroms of the crystallographic structure and the percentage of alignment coverage (**Figure 2**). The lower the resolution of a structure, the better its quality. The average resolution of the structures available in the PDB (Protein Data Bank) is around 3.5 Ä, while structures below 2.0 Ä are considered to have excellent resolution and represent less than 10% of the entries in the PDB. The higher the percentage of coverage of the alignment between a target protein (protein to be modeled) and the template (mold), the better [4]. Coverage alignments above 90% of the residues tends to have high scores and are considered to be excellent (**Figure 2**).

Something important to note in alignments is the presence of sequence gaps. A gap between sequences means the absence of residues, that is, amino acids that

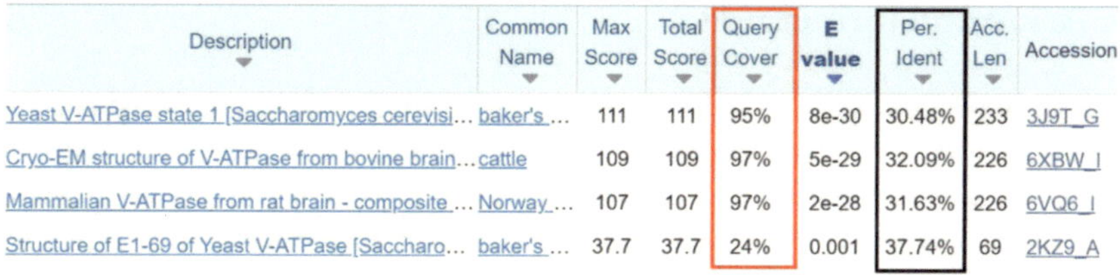

Description	Common Name	Max Score	Total Score	Query Cover	E value	Per. Ident	Acc. Len	Accession
Yeast V-ATPase state 1 [Saccharomyces cerevisi...	baker's ...	111	111	95%	8e-30	30.48%	233	3J9T_G
Cryo-EM structure of V-ATPase from bovine brain...	cattle	109	109	97%	5e-29	32.09%	226	6XBW_I
Mammalian V-ATPase from rat brain - composite ...	Norway ...	107	107	97%	2e-28	31.63%	226	6VQ6_I
Structure of E1-69 of Yeast V-ATPase [Saccharo...	baker's ...	37.7	37.7	24%	0.001	37.74%	69	2KZ9_A

Figure 2.
Example of BLASTp alignment between a Leishmania infantum *ATP-synthase sequence against the PDB database. Values of the coverage percentage (red) and identity (black) of each alignment are highlighted. Source: Authors data.*

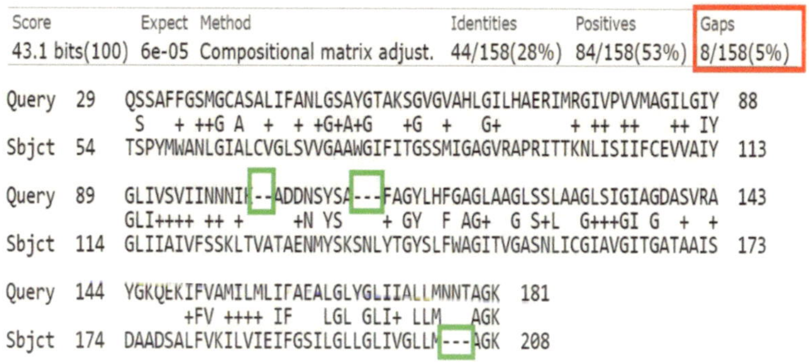

Figure 3.
Alignment between two proteins (query/Sbjct) showing the presence of 8 gaps (red) in three different sections (green). Source: Authors data.

have been deleted from some part of the sequence (**Figure 3**). The amount and size of gaps in an alignment is crucial to the final quality of the models. The greater the quantity and size of the gaps, the less reliable the models are and the greater is the chance of generating structural artifacts. Therefore, when choosing a template, it is essential that the researcher be aware about gaps presence in the sequences.

Once the template has been defined, we proceed to the stage of building the three-dimensional model. From specific programs and servers, the necessary files for modeling are submitted, which consists of the superimposition of the structural carbons of the target protein on the template protein, based on the alignment information to superimpose the equivalent amino acids. There are currently numerous free tools for building three-dimensional models (**Table 1**).

Nome	Tipo	Site
Modeler	Software	https://salilab.org/modeler/
Swiss-Model	Server	https://swissmodel.expasy.org/
Phyre2	Server	http://www.sbg.bio.ic.ac.uk/phyre2
Galaxy	Server	http://galaxy.seoklab.org/
RaptorX	Software/Server	http://raptorx.uchicago.edu/
CONFOLD	Software	https://github.com/multicom-toolbox/CONFOLD
ROBBETTA	Server	http://robetta.bakerlab.org/

Source: Google search.

Table 1.
Examples of free tools for building homology models.

2. Validation and refinement

Homology models are theoretical-computational approximations of the real protein structures, and therefore require validation and sometimes refinement and optimization. A very popular validation tool is the Ramachandran plot (**Figure 4**), which analyzes the stereochemical quality of protein structures.

The Ramachandran graph analyzes the conformations of the *phi* and *psi* angles of the peptide bonds, placing them in regions. Residues outside the permitted regions (outliers) are those that are in unfavorable configurations due to the collision between the atoms (steric shock). It preconizes that a good model should have at least 90% of its waste in favorable and permitted regions [5].

Other validation tools are energy assessments, both local and global ones. A tool for global assessment of the quality of a model is the server PROSA-web - Protein Structure Analysis (https://prosa.services.came.sbg.ac.at/prosa.php) [6, 7], which compares the energy of a structure with a database of proteins of equivalent size, solved experimentally, through the Z-score (**Figure 5**).

For local quality analysis, the application of the VERIFY3D server (https://servicesn.mbi.ucla.edu/Verify3D) is very useful. In this type of analysis it is possible to check the local quality, that is, for each residue of the model (**Figure 6**). With this, it is possible to identify specific regions of low quality for further adjustments.

For the models refinement, two techniques are particularly interesting: energy minimization and classical (atomistic) molecular dynamics. Energy minimization, also called optimization of geometry, aims to find a set of atomic coordinates of the structure that avoid bad contacts and reduce the potential energy of the system. There are some free servers available for energy minimization application in theoretical models, like YASARA [8] (http://www.yasara.org/minimizationserver.htm) and CHIRON [9] (https://dokhlab.med.psu.edu/chiron/). Molecular dynamics are extremely efficient for validating and refining theoretical models. This technique is based on the principles of Classical Mechanics and describes the atomic movements

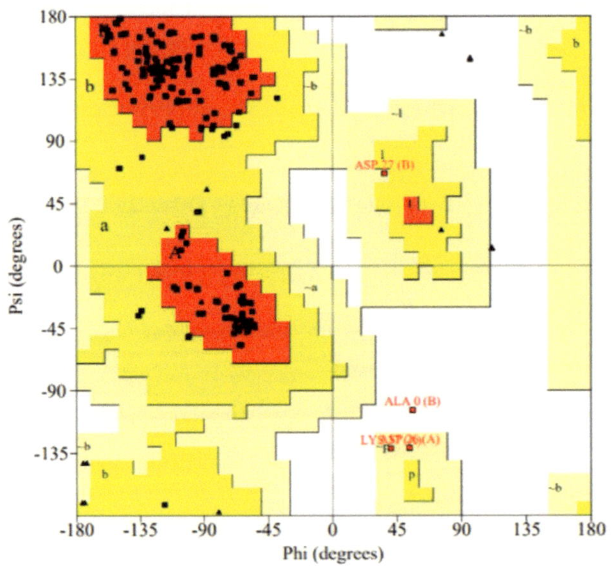

Figure 4.
Ramachandran graph for SARS-CoV-2 NSP9 replicase (PDB ID: 6w4b). In red, more favorable regions. In yellow and beige, regions allowed. In white, forbidden regions. Source: Authors data.

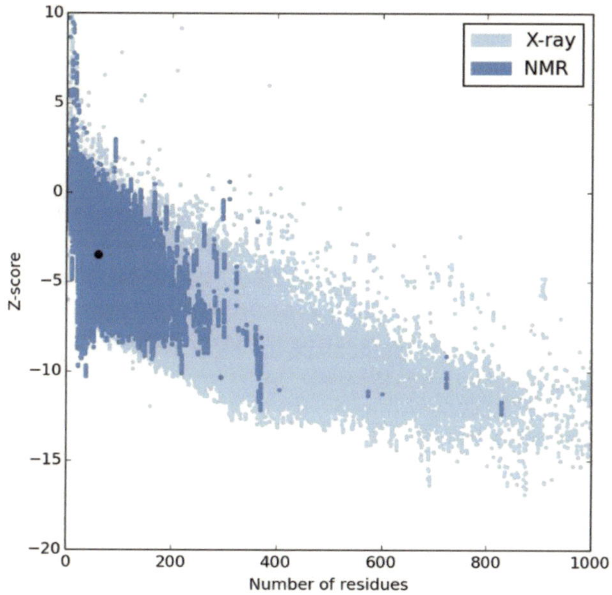

Figure 5.
Comparative graph of the Z-score energy. The black dot represents the position of the analyzed protein compared to equivalent size structures obtained by x-ray crystallography (light blue) and nuclear magnetic resonance (dark blue). Source: Authors data.

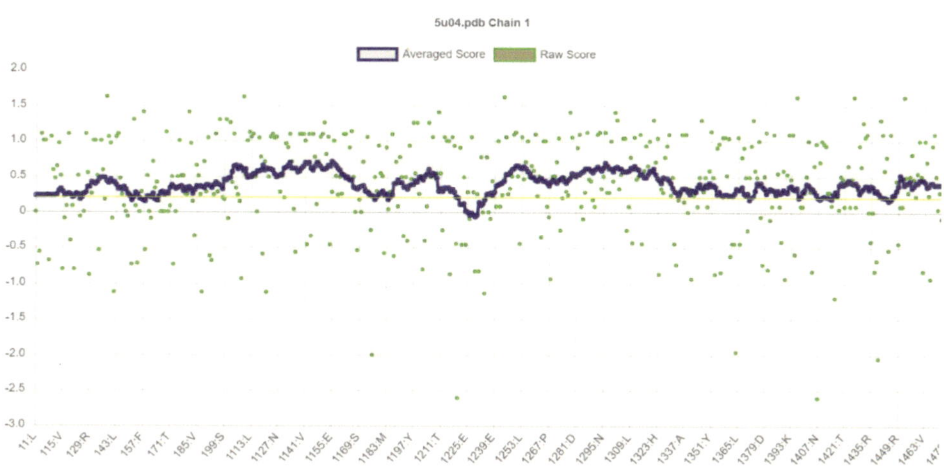

Figure 6.
Local ERRAT quality graph of a stretch from the NS5 enzyme from Zika virus. In blue, the average scores, in green, the raw scores. 93.93% of the residues have averaged 3D-1D score > = 0.2 (80% indicates good structures). Source: Authors data.

of a system through the integration of Newtonian equations of motion. Thus, a molecular dynamics simulation of 5–10 nanoseconds is one of the most effective techniques for optimization and validation of models by homology. For performing molecular dynamics calculations, software such as GROMACS [10] and NAMD [11] are useful. Once optimized and validated, the theoretical model can be used for several purposes, and can also be made available in public repositories, such as the PMDB - Protein Model DataBase (http://srv00.recas.ba.infn.it/PMDB/) and the SWISS-MODEL repository (https://swissmodel.expasy.org/repository).

3. Conclusions

Theoretical-computational models are fast, inexpensive and extremely versatile. There are countless possibilities for studies and uses of models by homology. These structures can be used for drug screening, docking studies, development of new drugs and vaccines, elucidation of binding sites (catalytic and allosteric), molecular dynamics simulations, quantum studies, biomolecule engineering etc.

The future of molecular modeling is fascinating and promising. With the advancement of computational tools, theoretical models tend to be increasingly accurate and reliable, contributing more and more to biological and biotechnological researches, in addition to integrating various areas of knowledge with bioinformatics and computational biology.

Acknowledgements

The authors are grateful to the Federal University of Campina Grande and to Federal Rural University of Pernambuco.

Conflict of interest

The authors declare no conflict of interest.

Author details

Rafael Trindade Maia[1*], Magnólia de Araújo Campos[2] and Rômulo Maciel de Moraes Filho[3]

1 Federal University of Campina Grande, Centro de Desenvolvimento Sustentável do Semiárido, Sumé, Paraíba State, Brazil

2 Federal University of Campina Grande, Centro de Educação e Saúde, Cuité, Paraíba State, Brazil

3 Federal Rural University of Pernambuco, Departamento de Agronomia, Recife, Pernambuco State, Brazil

*Address all correspondence to: rafael.rafatrin@gmail.com

References

[1] Luzzolino L, McGabe P, Prince S L, Brandenburg J G. Crystal structure prediction of flexible pharmaceutical-like molecules: density functional tight-binding as an intermediate optimisation method and for free energy estimation. Faraday Discuss., 2018, 211, 275.

[2] *Haddad Y*, *Adam V*, *Heger Z*. Ten quick tips for homology modeling of high-resolution protein 3D structures. *Plos Computational Biology*. *2020* 16(4): e1007449. doi:10.1371/journal.pcbi.1007449.

[3] Verli H. Bioinformática: da biologia à flexibilidade molecular. São Paulo, 2014. Ed. Sociedade Brasileira de Bioquímica e Biologia Molecular – SBBq.

[4] Leach AR. Molecular Modelling: Principles and Applications. Prentice Hall, 2001.

[5] Laskowski R A, MacArthur M W, Moss D S, Thornton J M (1993). PROCHECK - a program to check the stereochemical quality of protein structures. J. App. Cryst., 26, 283-291.

[6] Wiederstein, M. & Sippl, M.J. (2007) ProSA-web: interactive web service for the recognition of errors in three-dimensional structures of proteins. Nucleic Acids Research 35, W407-W410.

[7] Sippl, M.J. (1993). Recognition of Errors in Three-Dimensional Structures of Proteins.

[8] Krieger E, Joo K, Lee J, Lee J, Raman S, Thompson J, Tyka M, Baker D, Karplus K. Improving physical realism, stereochemistry, and side-chain accuracy in homology modeling: Four approaches that performed well in CASP8. Proteins. 2009;77 Suppl 9:114-22.

[9] Chiron : Ramachandran, S., Kota, P., Ding, F. and Dokholyan, N. V., PROTEINS: Structure, Function and Bioinformatics, 79: 261-270 (2011).

[10] Abraham, et al. (2015). GROMACS: High performance molecular simulations through multi-level parallelism from laptops to supercomputers. SoftwareX 1-2 19-25.

[11] James C. Phillips, David J. Hardy, Julio D. C. Maia, John E. Stone, Joao V. Ribeiro, Rafael *C. bernardi*, Ronak Buch, Giacomo Fiorin, Jerome Henin, Wei Jiang, Ryan McGreevy, Marcelo C. R. Melo, Brian K. Radak, Robert D. Skeel, Abhishek Singharoy, Yi Wang, Benoit Roux, Aleksei Aksimentiev, Zaida Luthey-Schulten, Laxmikant V. Kale, Klaus Schulten, Christophe Chipot, and Emad Tajkhorshid. Scalable molecular dynamics on CPU and GPU architectures with NAMD. Journal of Chemical Physics, 153:044130, 2020. doi:10.1063/5.0014475

Section 2

Reviews

Chapter 2

Normal Mode Analysis: A Tool for Better Understanding Protein Flexibility and Dynamics with Application to Homology Models

Jacob A. Bauer and Vladena Bauerová-Hlinková

Abstract

Molecular dynamics (MD) and normal mode analysis (NMA) are very useful methods for characterizing various dynamic aspects of biological macromolecules. In comparison to MD, NMA is computationally less expensive which facilitates the quick and systematic investigation of protein flexibility and dynamics even for large proteins and protein complexes, whose structure was obtained experimentally or *in silico*. In particular, NMA can be used to describe the flexible states adopted by a protein around an equilibrium position. These states have been repeatedly shown to have biological relevance and functional significance. This chapter briefly characterizes NMA and describes the elastic network model, a schematic model of protein shape used to decrease the computational cost of this method. Finally, we will describe the applications of this technique to several large proteins and their complexes as well as its use in enhancing protein homology modeling.

Keywords: normal mode analysis, elastic network model, crystal structure, protein dynamics, homology modeling

1. Introduction

Often there is a high demand for the structures of biologically important proteins, especially those which are large or part of complex systems. However, it is not always possible, for many reasons, to get a high-resolution structure experimentally using X-ray crystallography, NMR, or cryo-electron microscopy. Among the many problems, we can mention low protein expression, low protein stability, high aggregation or poorly diffracting crystals [1]. In this situation, *in silico* models provide a good starting point for experimental research. One of the common techniques for obtaining a reasonable structural model of a protein is homology modeling (HM). Homology modeling techniques are predominantly used to construct a hypothetical structure of a protein of interest (the target) where only the amino-acid sequence is available using the known structural features or 3D structure of one or several homologous proteins (the templates) [2–4]. However, building a static model often does not answer all questions regarding the function of the protein and its role in the cell and organism, nor does it clarify its relationship with other cellular components nucleic acids, proteins, ions and other molecules. It is important to understand protein dynamics and flexibility, as proteins during the fulfillment of their role in the cell often change their shape, oligomeric state or even fold. As it is often very challenging to observe protein dynamics *in vivo* or *in vitro,* to understand better all these changes, a useful computational method normal mode analysis (NMA) can be used [5–9].

2. Normal mode analysis

Normal mode analysis is a technique, based on the physics of small oscillations, that can be used to describe the flexible states accessible to a protein around an equilibrium position. The idea is that when an oscillating system at equilibrium, for example a protein in an energy minimum conformation, is slightly perturbed, a restoring force acts to bring the perturbed system back to its equilibrium conformation. A system is defined to be in equilibrium or at the bottom of a potential minimum when the generalized forces acting on it are equal to zero. At the minimum energy conformation, represented by the generalized coordinates q^0, the potential energy equation can be written as a power series in q:

$$V(q) = V(q^0) + \left(\frac{\partial V}{\partial q_i}\right)^0 \eta_i + \frac{1}{2}\left(\frac{\partial^2 V}{\partial q_i \partial q_j}\right)^0 \eta_i \eta_j + \ldots \tag{1}$$

where q_i and q_j are the instantaneous configuration of components i and j and the deviation of component i from its equilibrium configuration is given by $\eta_i = q_i - q_i^0$. $V(q)$ is the potential energy equation of the system and, for proteins, usually takes the form of one of the commonly used molecular dynamics force fields [10]. The first term in the series represents the minimum value of the potential and may be set to zero and the second term will be zero at any local minimum, so the potential can be written as

$$V(q) = \frac{1}{2}\left(\frac{\partial^2 V}{\partial q_i \partial q_j}\right)^0 \eta_i \eta_j = \frac{1}{2}\eta_i V_{ij} \eta_j \tag{2}$$

where V_{ij} is the Hessian matrix which contains the second derivatives of the potential with respect to the components of the system.

It is also necessary to consider the kinetic energy (T) of the system since we are interested in dynamics. For component i, this can be given by

$$T(q) = \frac{1}{2}M\frac{d^2\eta_i}{dt^2} \tag{3}$$

where M is a diagonal matrix containing the mass of each particle. The entire equation of motion can be written as

$$\frac{1}{2}M\frac{d^2\eta_i}{dt^2} + \frac{1}{2}\eta_i V_{ij} \eta_j = 0 \tag{4}$$

One solution of this equation is the oscillatory equation

$$\eta_i = a_{ik} \cos\left(\omega_k t + \delta_k\right) \tag{5}$$

where a_{ik} is the amplitude of oscillation, ω_k is the frequency, and δ_k is a phase factor. By substituting this into Eq. (4), the equation of motion can be rewritten as

$$VA = \lambda A \tag{6}$$

where the matrix A contains the A_k eigenvectors of the Hessian matrix V and λ is a diagonal matrix containing the λ_k eigenvalues. The A_k eigenvectors are the normal mode vectors and describe in which direction and how far each particle in the system moves with respect to each other particle; the λ_k eigenvalues give the squares of the frequencies with which all the particles involved with a particular mode vibrate. While the eigenvectors can tell in which direction and how far each particle moves with respect to the others, it does not give absolute displacements. NMA alone therefore cannot normally be used to get the displacement amplitudes of a given normal mode [11].

The vibrational energy of the system is generally equally divided so that every vibrational mode has the same energy and the average oscillation amplitude of a given mode scales as the inverse of its frequency. Thus, modes with higher frequencies, which will have energetically greater displacements, typically describe rapid but small amplitude local motions involving relatively few atoms, while those with lower frequencies will describe slower displacements involving larger numbers of atoms and describe large-scale conformational changes. As the name of the method indicates, these vibrational modes are normal to one another, meaning that they move independently: the excitation of one mode does not trigger the motion of a second one and the general motion of the system can be described by a superposition of all the modes. These normal modes yield analytical solutions to the equations of motion: for a given set of initial positions and velocities, NMA allows us to calculate where each atom of the system in question will be at any subsequent time subject to the small oscillation approximation. (A more complete treatment of the theory behind NMA and its advantages and limitations may be found in [12]).

NMA was first applied to peptides in 1979 [13] and was subsequently used to study the whole proteins bovine pancreatic trypsin inhibitor (BPTI) [14, 15], hexokinase [16], crambin [17], human lysozyme [17, 18], ribonuclease [17], and myoglobin [19, 20]. Application of the method to larger systems was hampered by its computational expense. With advances in computer technology and the development of more efficient algorithms, it has become possible to examine larger structures, including the skeletal ryanodine receptor [21], Ca-ATPase [22], GroEL [23–25], the ribosome [26–28], the yeast nuclear pore complex [29], and virus capsids [30–32]. All these will be examined in more detail in Section 3 below.

2.1 The elastic network model

NMA is less computationally expensive than molecular dynamics simulation, but it is still not trivial for proteins containing many thousands of atoms. The first problem is that the structure to be studied must be energy minimized to ensure that the starting conformation is in a true minimum relative to the chosen force field. The minimization must then proceed until machine precision is reached, typically below 0.001 kJ/mol-nm, which is much more computationally demanding than the minimizations normally employed for other tasks. Frequently, the results of this process distort the structure, leading to NMA being carried out on a structure different from the experimentally determined one. The second problem, and the computationally limiting factor, is the diagonalization of the Hessian matrix. For classical, all-atom NMA, all N atoms in a structure, including the hydrogen atoms, must be used, making the total Hessian $3N \times 3N$ in size. For large proteins with thousands of atoms this can become computationally difficult very quickly. Consequently, a number of coarse-grained approximate methods have been developed to overcome both of these limitations [6, 11]. The most common and widely used of these is the elastic network model (ENM).

The general idea of ENMs, first put forward by Tirion in 1996 [33], is to replace the complicated semi-empirical force fields used in standard NMA with a simple harmonic potential:

$$V(q) = \sum_{d_{ij} < R_c} C\left(d_{ij} - d_{ij}^0\right)^2 \quad (7)$$

where d_{ij} is the distance between atoms or nodes i and j, d_{ij}^0 is the distance in the initial structure, and C is a spring constant assumed to be the same for each i–j pair. It should be noted that by design the input configuration is assumed to be a minimum energy one, and energy minimization against a potential is therefore unnecessary. R_c in this equation refers to a cut-off radius and the sum is only over all pairs less than this value. R_c is somewhat arbitrary, but in practice, values of between 7.0–8.0 Å are used based on the observed distances between non-bonded atoms in experimental structures [34, 35]. Most frequently, only the C_α atoms are used for these calculations because they are sufficient for studying the backbone motions of the protein and are all that is necessary for characterizing the lowest-frequency normal modes.

A number of different ENM formulations have been developed. The simplest one is the Gaussian network model (GNM) developed by Bahar and co-workers [36]. The GNM replaces the $3N \times 3N$ Hessian matrix with an $N \times N$ Kirchoff matrix (Γ). Γ is defined in terms of spring constants γ_{ij}, which are created based on the assumption that the separation distance $|R_i - R_j| = R_{ij}$ between the ith and jth C_α atoms in the protein follows a Gaussian distribution. The potential is given by

$$V_{GNM} = \frac{1}{2}\sum_{ij}\gamma_{ij}\left(\Delta \overleftarrow{R_{ij}}\right)^2 \quad (8)$$

where $\Delta \overleftarrow{R_{ij}}$ is a vector expressing the fluctuations in distance between the ith and jth C_α atoms. The model assumes that these fluctuations are isotropic; consequently, no information about the three-dimensional directions of motion can be obtained. Eigenvalue decomposition of Γ does allow the contribution of individual modes to the equilibrium dynamics to be calculated, as well as the relative displacement of residues along each mode axis, the cross-correlation between the residues in the individual modes, and square displacement profiles.

Some form of the anisotropic network model (ANM) is perhaps the most commonly encountered ENM. This is the form originally suggested by Tirion [33] and incorporated into the Molecular Modeling Toolkit (MMTK) by Hinsen [37], and widely used in a number of other tools. The ANM gives the same information as the GNM, but also provides information on the directionality of the fluctuations. On the other hand, the mean-square fluctuations (*B*-factors) and cross-correlations it produces do not agree quite as well with experiment as GNM [38, 39]. In compensation, ANMs can be used to generate alternative conformations in the close neighborhood of the starting structure by deforming the structure along the lowest frequency modes [9]. Two groups led by Zheng [35] and Lin and Song [40] developed models which combined the best features of GNM and ANM into a single method.

While coarse-graining does allow ENMs to be scaled to very large models, it does so by losing detailed information on local structural movements. The rotating-translating blocks (RTB) model of Sanejouand [41] was constructed to alleviate this. In this approach, the protein or other macromolecule is divided into n_β blocks made up of one or a few residues connected by elastic springs. Next, it is assumed that a good approximation to the low-frequency normal modes can be made by forming linear combinations of the local rotations and translations of these individual blocks. Consequently, a $3N \times 6n_\beta$ projection matrix P is constructed and used to build a projected Hessian matrix H_β, which is diagonalized with $A_\beta^T H_\beta A_\beta = \Lambda_\beta$, where A_β is the eigenvector matrix diagonalizing H_β and Λ_β is the corresponding eigenvalue matrix. The resulting eigenvectors can be projected back into the full $3N$-dimensional space using $A_P = PA_\beta$, where A_P is a $3N \times 6n_\beta$ matrix containing the $6n_\beta$ lowest-frequency approximate normal modes.

3. Applications

In a survey of the recent structural biology literature, Bauer *et al.* [12] found that NMA, as a component of a structural study, most often was used for descrying the overall flexible motions of a molecule or for determining how those motions might be correlated with ligand binding or catalytic activation. Another use for NMA is to study the motions of macromolecules or macromolecular assemblies that are too large for treatment using conventional MD simulation. It has been shown that the low-frequency normal modes and the first few principal components calculated from a MD simulation overlap considerably, meaning that NMA can plausibly substitute for MD for large systems when only the overall collective motions of the system are needed [42–44]. Below, we will examine the application of NMA to experimentally obtained protein structures as well as to protein HM studies.

3.1 Application to experimental structures

3.1.1 Ryanodine receptor

The ryanodine receptors (RyRs) are the largest presently known ion channels, with molecular weights of 2.2 MDa. They are homotetramers embedded in the membrane of the sarcoplasmic reticulum of myocytes, where they play a key role in excitation-contraction coupling. They regulate Ca^{2+} release from the sarcoplasmic reticulum by undergoing a closed-to-open gating transition in response to an action potential or calcium binding. RyRs are found in all animals. Three isoforms have been identified in mammals: RyR1 (predominantly expressed in skeletal muscle), RyR2 (cardiac muscle) and RyR3 (present in several tissues including the brain, diaphragm, and testes) [45–47]. RyR malfunction leads to severe muscular disorders, including malignant hyperthermia, central core disease, tachycardia, dysplasia, and others [45]. The first high-resolution cryo-EM structures of the complete rabbit skeletal RyR were reported in 2015 [48–50], and were followed by a number of other high-resolution structures of the skeletal and cardiac isoforms in both their open and closed conformations, either alone or bound to regulators (For review see Bauerová-Hlinková *et al.* [51]).

A number of MD simulation studies have been reported for both the skeletal and cardiac RyR isoforms, but only covering small parts of the whole channel; in particular, the N-terminal domain (roughly the first 600 amino acids) [52–55] and parts of the channel domain [56–58] were studied in this way. Generally, these studies focused on identifying how known disease-causing mutations affected the dynamics of these fragments.

Shortly after the appearance of the first high-resolution cryo-EM structures, Wenjun Zheng published a normal-mode analysis on the 3.8-Å RyR1 structure [21]. He found that the largest collective motions of the closed form involved large outward and downward movements of the peripheral domains, in in good agreement with the conformational variations observed in multiple cryo-EM structures of the closed form later reported by des Georges *et al.* [59]. By using normal mode analysis to flexibly fit the closed form into a 10 Å cryo-EM map of the open form, he was also able to create a model of the open form of the receptor. A similar open-form conformation was created by Mowrey *et al.* [56] using an ion-pulling molecular dynamics simulation on only the transmembrane portion of the structure. Both models recreated the major features of the RyR1 and RyR2 open-channel structures reported subsequently, including the rotation of an important pore-lining transmembrane α-helix away from the channel axis.

3.1.2 Ca^{2+}-ATPase

The Ca^{2+}-ATPase may be thought of as the partner of the RyR. While RyR releases Ca^{2+} from the sarcoplasmic reticulum into the cytosol, the Ca^{2+}-ATPase pumps it back in, against a large concentration gradient, at the rate of two Ca^{2+} ions per hydrolyzed ATP. Conventionally, the Ca^{2+}-ATPase is thought to take two different states: E1, which has high affinity for Ca^{2+}, and E2, which has much lower affinity [60, 61]. In addition to the binding and dissociation of Ca^{2+}, ATP hydrolysis and dephosphorylation of the resulting phosphorylated Asp351 result in the presence of 4–7 different physiological states [62]. At least 54 crystal structures have been determined since the first one in 2000 [63] and they cover nearly all the conformations of the different physiological states. They show large conformational rearrangements during the reaction cycle and have been investigated by a number of computational methods [62]. Most of the NMA studies occurred soon after the initial structures by Toyoshima *et al.* [63, 64] had been reported, but before the final ones had become available [65, 66], making this a good situation for illustrating both the abilities and limitations of NMA.

The initial structure [63] showed that the ATPase consisted of three cytoplasmic domains, labeled A (activation), N (nucleotide binding), and P (phosphorylation), and 10 transmembrane helices (M1–M10). As the names suggest, the cytoplasmic domain N holds most of the ATP binding site and P holds the phosphorylation site. Subsequent structures [65, 66] showed that the A domain also participates in ATP binding and has a crucial role in dephosphorylation. In the first structure, taken to be in the E1 conformation, the N and A domains were widely separated. Subsequent E2 structures showed that ATP binding induces a large movement in the N domain and a smaller, though still considerable, motion in the A domain. NMA on only the initial E1 structure [67] found that the N domain seemed to be the most mobile and that rotational hinges were present between the N and P, and A and M domains. It was also suggested that the transmembrane α-helices M2, M4, and M8 likely played an important role in Ca^{2+} release into the sarcoplasmic reticulum lumen. After the second structure in the E2 conformation became available [64], two studies examined the normal modes that participate in the E1 $\rightarrow$ E2 conformational change [22, 68]. They found that only a few of the lower-frequency modes were needed to describe the E1 $\rightarrow$ E2 transition, which predominantly involved the movement of the A and N domains to close the cytoplasmic headpiece. They also both predicted that the transmembrane domain was likely to undergo a twisting motion which would eliminate the Ca^{2+} binding sites and open up the channel. The results of these studies were partially confirmed by the subsequent structures [65, 66]. The first

NMA study [67] did correctly predict that the largest conformational changes would be observed in the positions of the N and A domains, and the movement of helices M2, M4, and M8 was important for Ca^{2+} release, but they all missed that transmembrane helix M5 underwent a sharp bend following ADP release. They also failed to note that a conserved loop from domain A which interacts with the phosphate binding site shifts conformation and plays an important role in hydrolyzing the phosphate free from Asp351. Thus, these coarse-grained NMA studies managed to successfully predict many of the large-scale movements, but missed an important local conformational change and a large distortion of an element of secondary structure.

3.1.3 GroEL

Escherichia coli GroEL is one of the best studied chaperones and it is essential for cell viability [69, 70]. It is composed of two identical rings of seven subunits each. Each 548-residue subunit can be further subdivided into apical, intermediate, and equatorial domains, and the two rings stack back-to-back to form an isolated chamber where a non-native substrate can be refolded. The three domains are separated by upper and lower hinges, making each subunit highly flexible. GroEL works together with the co-chaperone GroES, which also has seven subunits and can bind to the apical part of either GroEL ring to form a cap [71]. During its activity cycle, GroEL goes through a number of conformational changes which are triggered by ATP binding and hydrolysis, interactions with GroES, and the substrate protein. When bound to no ligands, GroEL is in what is termed the "tense" or T state; this state is the most attractive to a (partially) unfolded substrate. Binding of ATP shifts GroEL to the "relaxed" or R state, and GroES binding and subsequent ATP hydrolysis shift it to the R″ state.

The structural transitions between these forms have been well described [70–75], and NMA has been applied to study the dynamics of both the individual subunits as well as the entire GroEL–GroES complex [5, 23–25, 76–78]. The earliest studies [23, 24] examined individual GroEL subunits and found that there is a very close relationship between their flexibility and the conformational changes observed for the entire complex. When ATP binds to a given subunit, the subunit changes conformation closely following a few low-frequency normal modes [23].

NMA was also calculated on an ENM of the whole GroEL–GroES complex [25]. These authors found that the slowest normal modes revealed a wide variety of motions which depended on the central cavity of the structure. These included the opposite twisting of the two GroEL rings combined with a flattening and expansion of the GroES cap; bending, shear, opposed radial breathing of the two rings, and stretching and contraction along the complex's long axis were also observed. They concluded that the mechanical motions driven by the different modes provide changing binding surfaces and differently sized cavities in the interior which might enable differently shaped substrates to be accommodated; possibly, these shifts might also be used during the refolding process.

3.1.4 The Ribosome

The ribosome is a molecular machine for translating the nucleotide sequence encoded in an mRNA transcript into a polypeptide sequence that folds into a functional protein. In prokaryotes, it is composed of a 50S subunit (containing a 23S rRNA, a 5S rRNA, and 34 proteins) and a 30S subunit (comprised of a 16S rRNA and 21 proteins) which together combine to form a 70S ribosome. The 30S subunit binds the mRNA and the anticodon end of the bound tRNA and is responsible for mRNA decoding. The 50S subunit interacts with the ends of the tRNA bound to the transferred amino acid and catalyzes peptide bond formation. The complete assembly features three sites for tRNA placement: an A (aminoacyl) site, a P (peptidyl) site, and an E (exit) site.

Several different crystal structures of ribosomes from different organisms and in different states have been determined by both X-ray crystallography and cryo-EM. The earliest were the 30S subunit from *Thermus thermophilus* at 3.0 Å [79, 80] and 3.3 Å resolution [81], the 50S subunit from *Haloarcula marismortui* at 2.4 Å [82, 83], and the complete *T. thermophilus* 70S ribosome at 5.5 Å [84]. These structures were the ones most commonly used for NMA studies. Many of these studies were carried out in the laboratory of Robert Jernigan [27, 85–89]. Overall, they found that the dominant motion was a ratchet motion between the large and small subunits, which was very similar to that observed experimentally by cryo-EM [26, 88]. This motion is observed regardless of whether the tRNAs and mRNA are present [87]. Curiously enough, this motion also remains even if all the proteins are stripped out, provided the general shape is maintained [27, 87]. This suggests that the motion arises from the ribosome structure itself and is not dependent upon the presence of its substrates.

Many other complex motions have been observed, especially at the mRNA decoding center in the 30S subunit [87]. In particular, the mRNA A site was found to be more flexible than the P site, which was consistent with the experimentally observed *B*-factors. It also agrees with the observation that the A site is able to accommodate a diverse set of substrates and that the P site needs to be more rigid to ensure the fidelity of the codon–anticodon matching. The collective dynamics of the exit tunnel for the growing polypeptide was also examined [28]. Here, it was found that the tunnel could be generally divided into three regions, entrance, neck, and exit, based on the low-frequency motions of the tunnel lining. Generally, the middle parts of the tunnel move in a complex way toward the tunnel exit, while the parts near the exit itself rotate around the tunnel axis. NMA of ENMs of the tRNAs were also examined and shown to be similar to the range of conformations observed from multiple experimental tRNA structures [89]. By comparing the normal modes of tRNAs alone and bound to the ribosome, they also noted that the ribosome acts to suppress all internal tRNA motions, only allowing it to move by rigid-body translation [90].

3.1.5 Yeast nuclear pore complex

The nuclear pore complex (NPC) is an enormous macroassembly that regulates the import and export of a large variety of substances (including proteins, nucleic acids, and small molecules) from the nucleus [91]. It is an octagonal complex composed of some 30 different proteins called nucleoporins. The yeast NPC has a mass of around 60 MDa while the vertebrate one is around 125 MDa. The yeast NPC is a ring that is around 100 nm across with a central pore of about 30 nm. The eight different subunits are termed "spokes" and each spoke exhibits pseudo 2-fold symmetry, giving the complex as a whole pseudo 16-fold symmetry. With 16 copies each, the 30 different nucleoporins make a total complex of 450–480 proteins. The central channel is coated with several "FG nucleoporins," which contain many structurally disordered phenylalanine (F) and glycine (G) repeats. These FG nucleoporins form a selective barrier: small particles ($<$ 30 kDa) can diffuse freely through the pore, while larger proteins require the assistance of karyopherin transport factors. The vertebrate NPC possesses, in addition to a central core, additional structural elements, including a cytoplasmic ring, a nuclear ring, and a luminal ring [91].

By following an integrative approach that combined data from many different sources, a coarse-grained structural model of the yeast NPC was developed [92, 93].

This model was then used for a coarse-grained NMA using an ENM [29]. This study found that two types of collective modes were predicted to be favored: a global bending mode and an extension and contraction mode that oscillated between a circle and an ellipse. Several different coarse-grained representations were tried and it was found that a simple model of a torid with an axial varying mass density was sufficient to capture the main dynamic features. It was also found that the number of spokes significantly affected the number of symmetric low-frequency modes: torids composed of eight spokes had access to symmetric modes that torids composed of seven or nine modes did not. A similar NMA study created using a different coarse-grained toridal model came to similar conclusions [94].

3.1.6 Virus Capsids

Aside from the ryanodine receptor and the nuclear pore complex, all of the forgoing examples are molecular machines of some sort that transform the chemical energy of ATP hydrolysis into mechanical motion. Viral capsids are different in that they are more of an architectural than a mechanical structure. Viral capsids encapsulate and protect the viral genome during its spread from cell to cell during infection. They can display a number of different surface features, including pores, canyons, spikes, and pillars. They typically consist of more than 100 protein subunits and have diameters of 100 nm or more. Capsids with more than 60 subunits that maintain their icosahedral symmetry are typically made up of collections of pentamers and hexamers [95]. The most commonly studied viral capsids form a spherical or icosahedral shape. NMA has been used to study the mechanical properties and conformational changes of these capsids for nearly 20 years (reviewed by May [96]). Even using coarse-grained NMA methods, the computational loads required to study whole capsids are still non-trivial, so a variety of symmetry-based approaches have been created to reduce them [32, 97–103].

One of the most thoroughly studied viral capsids is that of bacteriophage HK97 [31, 99, 104]. Tama and Brooks [31, 105] used NMA to study the swelling of a number of viral capsids, including cowpea chlorotic mottle virus (CCMV), Nudaurelia capensis virus (NωV), and HK97. They found that for CCMV and NωV, only the single lowest-frequency mode was needed to describe the swelling, while HK97 required the two lowest-frequency modes. The reason for this appeared to be that the CCMV and NωV, being spherical, required only a single mode, while HK97 not only expands, but also changes conformation from spherical to icosahedral. Generally, these studies, together with two others [106, 107], showed that the functional dynamics of the capsids are dictated by their structure, which is why only the lowest-frequency modes are needed. The HK97 expansion and conformational transition occurs during its maturation, and NMA has also been applied to study the HK97 maturation pathway [99, 104]. During maturation, a spherical procapsid undergoes substantial expansion together with a conformational change from spherical to icosahedral to form a mature capsid. These studies found that only a few low-frequency icosahedral modes were needed to account for most of the maturation expansion and that maturation appears to occur through the puckering of the pentamers followed by the flattening and cross-linking of the hexamers.

3.2 NMA in homology modeling studies

3.2.1 Membrane proteins

One of the protein groups in which homology modeling has been used together with NMA is multidomain transmembrane proteins, in particular the human nicotine acetylcholine receptor (hnAChR) [108] and *E. coli* mechanosensitive channel (MscL) [109]. hnAChRs are located mainly in the central nervous system and mediate fast neurotransmission. They possess a homopentameric quarternary structure where each monomer consists of an extracellular domain involving a conserved Cys loop and a ligand-binding channel, four transmembrane domains, and an intracellular segment. MscL is an integral membrane protein that gates in response to membrane tension in order to diminish the turgor pressure when bacteria are moved from a high to a low osmolarity environment. Like nAChR, MscL is a homopentamer where each subunit is composed of a cytoplasmic and a transmembrane domain.

Similar approaches were use in both studies. First a homology model of the target protein was constructed based on the known structures of similar proteins, followed by the application of NMA. The aim of both studies was to better understand channel gating and the particular structural changes associated with it [108, 109]. The hnAChR study also involved predicting the ligand binding site [108].

3.2.2 Enzymes

Enzymes have also been studied using a combination of HM and NMA. One example is the *Arabidopsis thaliana* Dicer-like 4 protein (AtDCL4) [110]. Dicer-like 4 is a large multidomain protein belonging to the Ribonuclease III family and is involved in the regulation of gene expression and antiviral defense through RNA-interference pathways. In particular, AtDCL4 produces short-stranded RNAs (ta-siRNA) which are incorporated into the RNA-induced silencing complex to direct the silencing of cognate RNA [111, 112]. The main aim of the study [110] was to better understand the mechanism of AtDCL4-mediated dsRNA recognition and binding by which small RNAs of a specific size are produced. First, the authors built the core of the AtDCL4 protein, which consists of a Platform, a PAZ domain, a Connector helix and RNAseA/B domains. A model of an AtDCL4–dsRNA complex was then constructed, which suggested that the spatial orientation of the AtDCL4 domains with respect to one another are responsible for the length of the bound dsRNA. Two regions, one on the surface of the Platform domain and second in the PAZ loop, were also identified, which are likely to be responsible for RNA binding.

3.2.3 Cell division and transport proteins

One of the longest-studied cellular processes is cell division [113] and cellular transport [114]. Regarding cell division, a combination of HM and NMA was used to study a yeast cohesin, an essential ring-shaped chromosome maintenance protein that mediates sister chromatid cohesion, homologous recombination, and DNA looping. This protein is a member of the structural maintenance of chromosomes (Smc) family, which exists in all eukaryotes [115]. In yeast, cohesin mainly consists of two Smc proteins, Smc1 and Smc3, both of which adopt long, anti-parallel coiled-coil regions that are separated by two globular regions: an ATP-binding head domain and a hinge region. The aim of the HM and NMA study [113] was to reveal the missing molecular details of how the two halves of the hinge region open to create an entry gate for DNA. In agreement with experimental data, the constructed yeast cohesin HM model showed that the bending motion of the cohesin ring is able to adopt a head-to-tail conformation. At the interface of the cohesin heterodimer, low-frequency conformational changes were observed to deform the highly conserved glycine residues present there. Normal mode analysis further revealed that the docking of large globular structures, such as the nucleosome and accessory proteins, to cohesin notably affected the mobility of the coiled-coil regions. Moreover, fully solvated molecular dynamics calculations, performed specifically on the hinge region, indicated that hinge opening starts from one side of the dimerization interface and is coordinated by the highly conserved glycine residues [113].

4. Conclusions

Normal mode analysis is a very useful technique for determining which conformational states are accessible to a given macromolecule. It can provide much of the same information given by more computationally expensive methods, such as molecular dynamics simulation, at only a fraction of the cost. It can be used by itself or in tandem with HM to characterize the general flexibility and domain movements of a molecule, to produce possible alternative conformations and confirm observed ones, and to describe the conformational changes that occur or might occur during substrate binding, product release, or catalytic activation. We have illustrated its utility using several examples of its application to a number of large biologically important proteins and protein complexes from bacteria, eukaryotes, and viruses. The biological relevance of the *in silico* models constructed by HM and NMA can be verified and expanded by different experimental approaches involving molecular and structural biology and biochemistry.

Acknowledgements

The authors would like to thank Eva Kutejová for general support during the writing of this chapter.

This research was funded by Vedecká Grantová Agentúra MŠVVaŠ SR and SAV grant number 2/0131/20.

Conflict of interest

The authors declare no conflict of interest.

Author details

Jacob A. Bauer* and Vladena Bauerová-Hlinková
Institute of Molecular Biology, Slovak Academy of Sciences, Bratislava, Slovakia

*Address all correspondence to: jacob.bauer@savba.sk

References

[1] McPherson A. Crystallization of Biological Macromolecules. Cold Spring Harbor, NY: Cold Spring Harbor Laboratory Press; 1999.

[2] Xiang Z. Advances in homology protein structure modeling. Curr Protein Pept Sci. 2006;7:217–227.

[3] Lohning AE, Levonis SM, Williams-Noonan B, Schweiker SS. A Practical Guide to Molecular Docking and Homology Modeling for Medicinal Chemists. Curr Top Med Chem. 2017;17: 2023–2040.

[4] Muhammed MT, Aki-Yalcin E. Homology modeling in drug discovery: Overview, current applications, and future perspectives. Chem Biol Drug Des. 2019;93:12–20.

[5] Ma J. Usefulness and Limitations of Normal Mode Analysis in Modeling Dynamics of Biomolecular Complexes. Structure. 2005;13:373–380.

[6] Skjaerven L, Hollup SM, Reuter N. Normal mode analysis for proteins. J Mol Struct Theochem. 2009;898:42–48.

[7] López-Blanco JR, Chacón P. New generation of elastic network models. Curr Opin Struct Biol. 2016;37:46–53.

[8] Tiwari SP, Reuter N. Conservation of intrinsic dynamics in proteins — what have computational models taught us? Curr Opin Struct Biol. 2018;50:75–81.

[9] Bahar I, Lezon TR, Bakan A, Shrivastava IH. Normal Mode Analysis of Biomolecular Structures: Functional Mechanisms of Membrane Proteins. Chem Rev. 2010;110:1463–1497.

[10] González MA. Force fields and molecular dynamics simulations. Collection SFN. 2011;12:169–200.

[11] Mahajan S, Sanejouand YH. On the relationship between low-frequency normal modes and the large-scale conformational changes of proteins. Arch Biochem Biophys. 2015;567:59–65.

[12] Bauer JA, Pavlovič J, Bauerová-Hlinková V. Normal Mode Analysis as a Routine Part of a Structural Investigation. Molecules. 2019;24:3293.

[13] Levy RM, Karplus M. Vibrational Approach to the Dynamics of an α-helix. Biopolymers. 1979;18:2465–2495.

[14] Noguti T, Gō N. Collective variable description of small-amplitude conformational fluctuations in a globular protein. Nature. 1982;296:776–778.

[15] Brooks B, Karplus M. Harmonic dynamics of proteins: Normal modes and fluctuations in bovine pancreatic trypsin inhibitor. Proc Natl Acad Sci USA. 1983;80:6571–6575.

[16] Harrison RW. Vibrational Calculation of the Normal Modes of a Large Macromolecule: Methods and some Initial Results. Biopolymers. 1984; 23:2943–2949.

[17] Levitt M, Sander C, Stern PS. Protein Normal-mode Dynamics: Trypsin Inhibitor, Crambin, Ribonuclease and Lysozyme. J Mol Biol. 1985;181:423–447.

[18] Brooks B, Karplus M. Normal modes for specific motions of macromolecules: Application to the hinge-bending mode of lysozyme. Proc Natl Acad Sci USA. 1985;82:4995–4999.

[19] Seno Y, Gō N. Deoxymyoglobin Studied by the Conformational Normal Mode Analysis I. Dynamics of Globin and the Heme–Globin Interaction. J Mol Biol. 1990;216:95–109.

[20] Seno Y, Gō N. Deoxymyoglobin Studied by the Conformational Normal

Mode Analysis II. The Conformational Change upon Oxygenation. J Mol Biol. 1990;216:111–126.

[21] Zheng W. Toward decrypting the allosteric mechanism of the ryanodine receptor based on coarse-grained structural and dynamic modeling. Proteins. 2015;83:2307–2318.

[22] Li G, Cui Q. Analysis of Functional Motions in Brownian Molecular Machines with an Efficient Block Normal Mode Approach: Myosin-II and Ca^{2+}-ATPase. Biophys J. 2004;86:743–763.

[23] Ma J, Karplus M. The allosteric mechanism of the chaperonin GroEL: A dynamic analysis. Proc Natl Acad Sci USA. 1998;95:8502–8507.

[24] Ma J, Sigler PB, Xu Z, Karplus M. A Dynamic Model for the Allosteric Mechanism of GroEL. J Mol Biol. 2000; 302:303–313.

[25] Keskin O, Bahar I, Flatow D, Covell DG, Jernigan RL. Molecular Mechanisms of Chaperonin GroEL–GroES Function. Biochemistry. 2002;41: 491–501.

[26] Tama F, Valle M, Frank J, Brooks III CL. Dynamic reorganization of the functionally active ribosome explored by normal mode analysis and cryo-electron microscopy. Proc Natl Acad Sci USA. 2003;100:9319–9323.

[27] Yan A, Wang Y, Klochkowski A, Jernigan RL. Effects of Protein Subunits Removal on the Computed Motions of Partial 30S Structures of the Ribosome. J Chem Theory Comput. 2008;4:1757–1767.

[28] Kurkcuoglu O, Kurkcuoglu Z, Doruker P, Jernigan RL. Collective dynamics of the ribosomal tunnel revealed by elastic network modeling. Proteins. 2009;75:837–845.

[29] Lezon TR, Sali A, Bahar I. Global Motions of the Nuclear Pore Complex: Insights from Elastic Network Models. PLoS Comput Biol. 2009;5:e1000496.

[30] Lee BH, Jo S, Choi Mk, Kim MH, Choi JB, Kim MK. Normal mode analysis of Zika virus. Comput Biol Chem. 2018; 72:53–61.

[31] Tama F, Brooks III CL. Diversity and Identity of Mechanical Properties of Icosahedral Viral Capsids Studied with Elastic Network Normal Mode Analysis. J Mol Biol. 2005;345:299–314.

[32] van Vlijmen HWT, Karplus M. Normal Mode Calculations of Icosahedral Viruses with Full Dihedral Flexibility by Use of Molecular Symmetry. J Mol Biol. 2005;350:528–542.

[33] Tirion MM. Large Amplitude Elastic Motions in Proteins from a Single-Parameter, Atomic Analysis. Phys Rev. Lett. 1996;77:1905–1908.

[34] Miyazawa S, Jernigan RL. Estimation of Effective Interresidue Contact Energies from Protein Crystal Structures: Quasi-Chemical Approximation. Macromolecules. 1985;18:534–552.

[35] Zheng W. A Unification of the Elastic Network Model and the Gaussian Network Model for Optimal Description of Protein Conformational Motions and Fluctuations. Biophys J. 2008;94: 3853–3857.

[36] Bahar I, Atilgan AR, Erman B. Direct evaluation of thermal fluctuations in proteins using a single-parameter harmonic potential. Folding Des. 1997;2:173–181.

[37] Hinsen K. The Molecular Modeling Toolkit: A New Approach to Molecular Simulations. J Comp Chem. 2000;21: 79–85.

[38] Bahar I, Rader AJ. Coarse-grained normal mode analysis in structural biology. Curr Opin Struct Biol. 2005;15: 586–592.

[39] Cui Q, Bahar I, editors. Normal Mode Analysis. Theory and Applications to Biological and Chemical Systems. Boca Raton, Florida, USA: Chapman & Hall/CRC; 2006.

[40] Lin TL, Song G. Generalized spring tensor models for protein fluctuation dynamics and conformation changes. BMC Struct Biol. 2010;10(Suppl 1):S3.

[41] Durand P, Trinquier G, Sanejouand YH. A New Approach for Determining Low-Frequency Normal Modes in Macromolecules. Biopolymers. 1994;34:759–771.

[42] Rueda M, Chacón P, Orozco M. Thorough Validation of Protein Normal Mode Analysis: A Comparative Study with Essential Dynamics. Structure. 2007;15:565–575.

[43] Rueda M, Ferrer-Costa C, Meyer T, Pérez A, Camps J, Hospital A, et al. A consensus view of protein dynamics. Proc Natl Acad Sci USA. 2007;104:796–801.

[44] Doruker P, Atilgan R Ali, Bahar I. Dynamics of Proteins Predicted by Molecular Dynamics Simulations and Analytical Approaches: Application to α-Amylase Inhibitor. Proteins. 2000;40: 512–524.

[45] Lanner JT, Georgiou DK, Joshi AD, Hamilton SL. Ryanodine Receptors: Structure, Expression, Molecular Details, and Function in Calcium Release. Cold Spring Harb Perspect Biol. 2010;2:a003996.

[46] Van Petegem F. Ryanodine Receptors: Structure and Function. J Biol Chem. 2012;287:31624–31,632.

[47] Van Petegem F. Ryanodine Receptors: Allosteric Ion Channel Giants. J Mol Biol. 2015;427:31–53.

[48] Zalk R, Clarke OB, des Georges A, Grassucci RA, Reiken S, Mancia F, et al. Structure of a mammalian ryanodine receptor. Nature. 2015;517:44–49.

[49] Yan Z, Bai Xc, Yan C, Wu J, Li Z, Xie T, et al. Structure of the rabbit ryanodine receptor RyR1 at near-atomic resolution. Nature. 2015;517:50–55.

[50] Efremov RG, Leitner A, Aebersold R, Raunser S. Architecture and Conformational Switch Mechanism of the Ryanodine Receptor. Nature. 2015;517:39–43.

[51] Bauerová-Hlinková V, Hajdúchová D,, Bauer JA. Structure and Function of the Human Ryanodine Receptors and Their Association with Myopathies – Present State, Challenges, and Perspectives. Molecules. 2020;.

[52] Bauer JA, Borko v, Pavlović J, Kutejová E, Bauerová-Hlinková V. Disease-associated mutations alter the dynamic motion of the N-terminal domain of the human cardiac ryanodine receptor. J Biomol Struct Dyn. 2020;38: 1054–1070.

[53] Yamazawa T, Ogawa H, Murayama T, Yamaguchi M, Oyamada H, Suzuki J, et al. Insights into channel modulation mechanism of RYR1 mutants using Ca^{2+} imaging and molecular dynamics. J Gen Physiol. 2020;152:e201812235.

[54] Zheng W, Liu Z. Investigating the inter-subunit/subdomain interactions and motions relevant to disease mutations in the N-terminal domain of ryanodine receptors by molecular dynamics simulation. Proteins. 2017;85: 1633–1644.

[55] Xiong J, Liu X, Gong Y, Zhang P, Qiang S, Zhao Q, et al. Pathogenic mechanism of a catecholaminergic polymorphic ventricular tachycardia causing-mutation in cardiac calcium release channel RyR2. J Mol Cell Cardiol. 2018;117:26–35.

[56] Mowrey DD, Xu L, Mei Y, Pasek DA, Meissner G,, et al. Ion-pulling simulations provide insights into the

mechanisms of channel opening of the skeletal muscle ryanodine receptor. J Biol Chem. 2017;292:12947–12,958.

[57] Schilling R, Fink RHA, Fischer WB. MD simulations of the central pore of ryanodine receptors and sequence comparison with 2B protein from coxsackie virus. Biochim Biophys Acta. 2014;1838:1122–1131.

[58] Schilling R, Fink RHA, Fischer WB. Interaction of ions with the luminal sides of wild-type and mutated skeletal muscle ryanodine receptors. J Mol Model. 2016;22:37.

[59] des Georges A, Clarke OB, Zalk R, Yuan Q, Condon KJ, Grassucci RA, et al. Structural basis for gating and activation of RyR1. Cell. 2016;167:145–157.

[60] Albers RW. Biochemical aspects of active transport. Annu Rev. Biochem. 1967;36:727–756.

[61] Post RL, Hegyvary C, Kume S. Activation by Adenosine Triphosphate in the Phosphorylation Kinetics of Sodium and Potassium Ion Transport Adenosine Triphosphatase. J Biol Chem. 1972;247:6530–6540.

[62] Kobayashi C, Koike R, Ota M, Sugita Y. Hierarchical domain-motion analysis of conformational changes in sarcoplasmic reticulum Ca^{2+}-ATPase. Proteins. 2015;83:746–756.

[63] Toyoshima C, Nakasako M, Nomura H, Ogawa H. Crystal structure of the calcium pump of sarcoplasmic reticulum at 2.6 A° resolution. Nature. 2000;405:647–655.

[64] Toyoshima C, Nomura H. Structural changes in the calcium pump accompanying the dissociation of calcium. Nature. 2002;418:605–611.

[65] Toyoshima C, Mizutani T. Crystal structure of the calcium pump with a bound ATP analogue. Nature. 2004;430: 529–535.

[66] Toyoshima C, Nomura H, Tsuda T. Lumenal gating mechanism revealed in calcium pump crystal structures with phosphate analogues. Nature. 2004;432: 361–368.

[67] Li G, Cui Q. A Coarse-Grained Normal Mode Approach for Macromolecules: An Efficient Implementation and Application to Ca^{2+} -ATPase. Biophys J. 2002;83:2457–2474.

[68] Reuter N, Hinsen K, Lacap'ere JJ. Transconformations of the SERCA1 Ca-ATPase: A Normal Mode Study. Biophys J. 2003;85:2186–2197.

[69] Thirumalai D, Lorimer GH. Chaperonin-Mediated Protein Folding. Annu Rev. Biophys Biomol Struct. 2001; 30:245–269.

[70] Horwich AL. Chaperonin-Mediated Protein Folding. J Biol Chem. 2013;288: 23622–23,632.

[71] Xu Z, Horwich AL, Sigler PB. The crystal structure of the asymmetric GroEL–GroES–$(ADP)_7$ chaperonin complex. Nature. 1997;388:741–750.

[72] Chaudhry C, Horwich AL, Brunger AT, Adams PD. Exploring the Structural Dynamics of the *E. coli* Chaperonin GroEL Using Translation-libration-screw Crystallographic Refinement of Intermediate States. J Mol Biol. 2004;342:229–245.

[73] Hyeon C, Lorimer GH, Thirumalai D. Dynamics of allosteric transitions in GroEL. Proc Natl Acad Sci USA. 2006;103:18939–18,944.

[74] Liu J, Sankar K, Wang Y, Jia K, Jernigan RL. Directional Force Originating from ATP Hydrolysis Drives the GroEL Conformational Change. Biophys J. 2017;112:1561–1570.

[75] de Groot BL, Vriend G, Berendsen HJC. Conformational Changes in the Chaperonin GroEL: New Insights Into the Allosteric Mechanism. J Mol Biol. 1999;286:1241–1249.

[76] Yang Z, Májek P, Bahar I. Allosteric Transitions of Supramolecular Systems Explored by Network Models: Application to Chaperonin GroEL. PLoS Comput Biol. 2009;5:e1000360.

[77] Na H, Jernigan RL, Song G. Bridging between NMA and Elastic Network Models: Preserving All-Atom Accuracy in Coarse-Grained Models. PLoS Comput Biol. 2015;11:e1004542.

[78] Zheng W, Brooks BR, Thirumalai D. Allosteric Transitions in the Chaperonin GroEL are Captured by a Dominant Normal Mode that is Most Robust to Sequence Variations. Biophys J. 2007;93: 2289–2299.

[79] Wimberly BT, Brodersen DE, Clemons Jr. WM, Morgan-Warren RJ, Carter AP, Vonrhein C, et al. Structure of the 30S ribosomal subunit. Nature. 2000;407:327–339.

[80] Carter AP, Clemons WM, Brodersen DE, Morgan-Warren RJ, Wimberly BT, Ramakrishnan V. Functional insights from the structure of the 30S ribosomal subunit and its interactions with antibiotics. Nature. 2000;407:340–348.

[81] Schluenzen F, Tocilj A, Zarivach R, Harms J, Gluehmann M, Janell D, et al. Structure of Functionally Activated Small Ribosomal Subunit at 3.3 Å Resolution. Cell. 2000;102:615–623.

[82] Ban N, Nissen P, Hansen J, Moore PB, Steitz TA. The Complete Atomic Structure of the Large Ribosomal Subunit at 2.4 Å Resolution. Science. 2000;289:905–920.

[83] Nissen P, Hansen J, Ban N, Moore PB, Steitz TA. The Structural Basis of Ribosome Activity in Peptide Bond Synthesis. Science. 2000;289: 920–930.

[84] Yusupov MM, Yusupova GZ, Baucom A, Lieberman K, Earnest TN, Cate JHD, et al. Crystal Structure of the Ribosome at 5.5 Å Resolution. Science. 2001;292:883–896.

[85] Zimmermann MT, Jia K, Jernigan RL. Ribosome Mechanics Informs about Mechanism. J Mol Biol. 2016;428:802–810.

[86] Kurkcuoglu O, Turgut OT, Cansu S, Jernigan RL, Doruker P. Focused Functional Dynamics of Supramolecules by Use of a Mixed-Resolution Elastic Network Model. Biophys J. 2009;97: 1178–1187.

[87] Kurkcuoglu O, Doruker P, Sen TZ, Kloczkowski A, Jernigan RL. The ribosome structure controls and directs mRNA entry, translocation and exit dynamics. Phys Biol. 2008;5:046005.

[88] Wang Y, Rader AJ, Bahar I, Jernigan RL. Global ribosome motions revealed with elastic network model. J Struct Biol. 2004;147:302–314.

[89] Zimmermann MT, Jernigan RL. Elastic network models capture the motions apparent within ensembles of RNA structures. RNA. 2014;20:792–804.

[90] Wang Y, Jernigan RL. Comparison of tRNA Motions in the Free and Ribosomal Bound Structures. Biophys J. 2005;89:3399–3409.

[91] Wente SR, Rout MP. The Nuclear Pore Complex and Nuclear Transport. Cold Spring Harb Perspect Biol. 2010;2: a000562.

[92] Alber F, Dokudovskaya S, Veenhoff LM, Zhang W, Kipper J, Devos D, et al. Determining the architectures of macromolecular assemblies. Nature. 2007;450:683–694.

[93] Alber F, Dokudovskaya S, Veenhoff LM, Zhang W, Kipper J, Devos D, et al. The molecular architecture of the nuclear pore complex. Nature. 2007;450:695–701.

[94] Wolf C, Mofrad MRK. On the Octagonal Structure of the Nuclear Pore Complex: Insights from Coarse-Grained Models. Biophys J. 2008;95:2073–2085.

[95] Caspar DLD, Klug A. Physical Principles in the Construction of Regular Viruses. Cold Spring Harb Symp Quant Biol. 1962;27:1–24.

[96] May ER. Recent Developments in Molecular Simulation Approaches to Study Spherical Virus Capsids. Mol Simul. 2014;40:878–888.

[97] van Vlijmen HWT, Karplus M. Normal mode analysis of large systems with icosahedral symmetry: Application to (Dialanine)$_{60}$ in full and reduced basis set implementations. J Chem Phys. 2001;115:691–698.

[98] Ming D, Kong Y, Wu Y, Ma J. Substructure synthesis method for simulating large molecular complexes. Proc Natl Acad Sci USA. 2003;100: 104–109.

[99] Kim MK, Jernigan RL, Chirikjian GS. An elastic network model of HK97 capsid maturation. J Struct Biol. 2003;143:107–117.

[100] Dykeman EC, Sankey OF. Atomistic modeling of the low-frequency mechanical modes and Raman spectra of icosahedral virus capsids. Phys Rev. E Stat Nonlin Soft Matter Phys. 2010;81:021918.

[101] Chen X, Sun Y, An X, Ming D. Virtual interface substructure synthesis method for normal mode analysis of super-large molecular complexes at atomic resolution. J Chem Phys. 2011;135:144108.

[102] Lu M, Ming D, Ma J. fSUB: Normal Mode Analysis with Flexible Substructures. J Phys Chem B. 2012;116: 8636–8645.

[103] Martín-Bravo M, Llorente JMG, Hernández-Rojas J. A minimal coarse-grained model for the low-frequency normal mode analysis of icosahedral viral capsids. Soft Matter. 2020;16: 3443–3455.

[104] Rader AJ, Vlad DH, Bahar I. Maturation Dynamics of Bacteriophage HK97 Capsid. Structure. 2005;13:413–421.

[105] Tama F, Brooks III CL. The Mechanism and Pathway of pH Induced Swelling in Cowpea Chlorotic Mottle Virus. J Mol Biol. 2002;318:733–747.

[106] Englert F, Peeters K, Taormina A. Twenty-four near-instabilities of Caspar-Klug viruses. Phys Rev. E Stat Nonlin Soft Matter Phys. 2008;78: 031908.

[107] Peeters K, Taormina A. Group theory of icosahedral virus capsid vibrations: A top-down approach. J Theor Biol. 2009;256:607–624.

[108] Haddadian EJ, Cheng MH, Coalson RD, Xu Y, Tang P. In Silico models for the Human $\alpha4\beta2$ Nicotinic Acetylcholine Receptor. J Phys Chem B. 2008;112:13981–13,990.

[109] Valadié H, Lacapčre JJ, Sanejouand YH, Etchebest C. Dynamical Properties of the MscL of *Escherichia coli*: A Normal Mode Analysis. J Mol Biol. 2003;332:657–674.

[110] Mickiewicz A, Sarzyńska J, Miłostan M, Kurzyńska-Kokorniak A, Rybarczyk A, Łukasiak P, et al. Modeling of the catalytic core of *Arabidopsis thaliana* Dicer-like 4 protein and its complex with double-stranded RNA. Comput Biol Chem. 2017;66:44–56.

[111] Brodersen P, Sakvarelidze-Achard L, Bruun-Rasmussen M, Dunoyer P,

Yamamoto YY, Sieburth L, et al. Widespread Translational Inhibition by Plant miRNAs and siRNAs. Science. 2008;320:1185–1190.

[112] Vazquez F, Hohn T. Biogenesis and Biological Activity of Secondary siRNAs in Plants. Scientifica. 2013;2013:783253.

[113] Kurkcuoglu O, Bates PA. Mechanism of Cohesin Loading onto Chromosomes: A Conformational Dynamics Study. Biophys J. 2010;99: 1212–1220.

[114] Serohijos AWR, Chen Y, Ding F, Elston TC, Dokholyan NV. A structural model reveals energy transduction in dynein. Proc Natl Acad Sci USA. 2006; 103:18540–18,545.

[115] Nasmyth K, Haering CH. Cohesin: its roles and mechanisms. Annu Rev. Genet. 2009;43:525–558.

Chapter 3

Role of Force Fields in Protein Function Prediction

Zaved Hazarika, Sanchaita Rajkhowa and Anupam Nath Jha

Abstract

The world today, although, has developed an elaborate health system to fortify against known and unknown diseases, it continues to be challenged by new as well as emerging, and re-emerging infectious disease threats with severity and probable fluctuations. These threats also have varying costs for morbidity and mortality, as well as for a complex set of socio-economic outcomes. Some of these diseases are often caused by pathogens which use humans as host. In such cases, it becomes paramount responsibility to dig out the source of pathogen survival to stop their population growth. Sequencing genomes has been finessed so much in the 21st century that complete genomes of any pathogen can be sequenced in a matter of days following which; different potential drug targets are needed to be identified. Structure modeling of the selected sequences is an initial step in structure-based drug design (SBDD). Dynamical study of predicted models provides a stable target structure. Results of these *in-silico* techniques greatly depend on force field (FF) parameters used. Thus, in this chapter, we intend to discuss the role of FF parameters used in protein structure prediction and molecular dynamics simulation to provide a brief overview on this area.

Keywords: homology modeling, force field (FF), molecular dynamics (MD) simulations, molecular docking

1. Introduction

What is a "disease"? A disease is any condition that harms the normal function of a body organ and/or system, of the psyche, or of the organism as a whole, which is associated with specific signs and symptoms. Factors that often lead to the damage of the function of organs and/or systems may be of two types, i.e., intrinsic and extrinsic. Those factors, that arise from within the host body interfering with the normal functioning processes of a body organ and/or system, as a result of genetic features of an organism or any disorder within the host are known as intrinsic factors [1]. Huntington's disease is an example of genetic disease which causes uncontrolled movements, emotional problems and loss of thinking ability (cognition) owing to a progressive brain disorder, due to mutations in the *HTT* gene, involving a DNA segment known as CAG trinucleotide repeats [2]. When a host comes in contact with a pathogen from outside, the host's system is accessed by extrinsic factors [3]. Microorganisms are the main causative agents which are responsible for causing infectious diseases. Their importance is determined from the type and extent of damage their causative agents inflict on organs and/or systems when they enter into a host. Entry into the host is mostly by routes such as the mouth, eyes, genital openings, nose and the skin. Damage to tissues mainly results from the growth and metabolic processes of infectious agents intracellular or within body fluids, with the production and release of toxins or enzymes that interfere with the normal functioning of organs and/or systems [4]. An example of extrinsic factor is the infection caused by novel pathogen, such as SARS-CoV-2, which represents an extremely challenging and complex endeavor. Currently, several promising therapeutics are underway and also many vaccine candidates with promises to mitigate the catastrophic effects of COVID-19 pandemic are under clinical trials. Still, an effective and successful countermeasure to control this catastrophe is not available [5].

In December 2019, a kind of pneumonia having an unknown etiology was reported from the Wuhan city of China in the Hubei province [6]. Isolation and genomic characterization of the complete sequence of the virus using next-generation sequencing (NGS), identified it as a novel coronavirus (CoV) and named it as 2019-nCoV, now as SARS-CoV-2 [7]. Although the characterization of the complete sequence was completed in January 2020, yet till date, there is no definitive cure or vaccine available for this virus. With the availability of the sequence, the three-dimensional (3D) structures of many proteins belonging to SARS-CoV-2 are now available. These 3D-structures can be obtained using various experimental and computational techniques. X-ray crystallography and NMR spectroscopy are currently the two major experimental techniques for protein structure determination [8] which are deposited in both UniProt and Protein Data Bank (PDB) [9]. For computational modeling of the 3D structure of proteins, homology modeling technique is used. Homology modeling is a computational technique which uses the amino acid sequence to predict the 3D structure. It is one of the widely used computational structure prediction method.

Proteins are one of the most extensively studied and complex macromolecules within living organisms with a unique 3D structure. Usually this leads to a diversity in their spatial shape, structure and thus, leading to different biological functionalities in a living system [8]. Yet, very little is known about the process of protein folding leading to its specific tertiary structure from its primary structure. Till date, approximately 175,000 experimentally determined 3D structures of biological macromolecules are available in the PDB [9]. However, reference sequence (refseq) release of National Center for Biotechnology Information (NCBI) contains as many as 178,304,046 protein sequences. This signifies a huge difference between the number of sequences in the NCBI and the number of protein 3D structures in the PDB. The difference in the number is even higher due to the fact that the reference sequences in the NCBI are non-redundant, whereas, structures available in PDB contain redundancy. This has resulted in an alarming situation owing to the increasing gap between the available 3D structures and the protein sequences. Therefore, computational structural prediction methods such as homology modeling are much needed in covering this widening gap. Thus, this chapter discusses homology modeling in a holistic manner covering the principles and different types of structure prediction methods along with giving a flavor of the different force field (FF) parameters that are used in protein structure prediction. The chapter also includes a brief overview of the molecular dynamics (MD) simulations that are used in computational modeling of proteins along with discussion of some application examples in this field.

2. Protein structure prediction

Protein sequences are much easier to obtain as compare to their structures. This is due to advancements in the field of protein sequencing technology. As a result, an exponential growth in the accumulation of protein sequences can be observed. An amino acid sequence is a very important source of insight into proteins, its function, structure and history. This is mostly because, first, comparison of an unknown sequence with a known sequence helps in deciding whether significant similarities exist between them, which in turn helps in establishing the class of protein and can give valuable information regarding its structure and function. Secondly, genealogical relationships can be studied by comparing the sequences of the same protein from different species. Thirdly, the presence of internal repeats in protein sequences reveals the history of the proteins. Also, sequencing of amino acids is very important for making DNA probes which can be used for encoding of its protein, as knowledge of the primary structure also allows the use of reverse genetics [10].

2.1 Amino acid sequence determination techniques

Determination of the amino acid sequence of all or part of a protein or peptide is known as prediction of protein sequence. It is used to categorize the protein and may help in characterizing its post-translational modifications. In a protein, determination of the amino acid sequence involves the following steps [10]:

i. *Hydrolysis*: This procedure is required in order to hydrolyze the protein into its amino acid and includes the protein being heated in 6 M hydrochloric acid (HCl) at 100–110° C for 24 hours or longer.

ii. *Separation*: Separation of amino acid from a peptide can be achieved by ion-exchange chromatography. The amino acids are eluted by mixing them with an acidic solution and passing a buffer steadily while increasing the pH through the chromatography column on sulfonated polystyrene. Accordingly, when an amino acid reaches its isoelectric point, it is separated. The buffer used is correlated to a specific amino acid type. Thus, the amino acid having the most acidic side chain will emerge first, while the amino acid having the most basic side chain will emerge last. The absorbance is used to determine the amount of similar type amino acid residues.

iii. *Quantitation*: Once the separation of the amino acids is achieved, their respective quantities are determined by adding a reagent called ninhydrin which gives an intense blue color to the amino acids, except proline which, due to the presence of secondary amino group in its structure, gives it a yellow color. For very small quantities (nanogram), reagents like fluorescamine or ortho-phthaldehyde (OPA) are used to obtain fluorescent products. Therefore, the concentration of amino acids is directly proportional to either the absorbance of the resulting solution or the fluorescence emitted by the sample.

For determining the composition and the sequence of the protein, two direct methods can be used:

a. *Edward Degradation Method*: This method uses phenyl iso-thio-cyanate to cleave the amino acids one by one starting from the amino terminal. The amino acids when treated with phenyl iso-thio-cyanate forms a phenyl-thio-hydantoin (PTH)-amino acid (e.g. PTH-lysine, etc.) terminal residue, which gets released under mild acidic conditions. The released terminal compound is then identified using chromatographic procedures.

b. *Mass Spectrometry*: Another technique to determine protein sequence is the mass spectrometry which uses the time of flight of ionized proteins to calculate the mass of the ionized proteins. In this process, the protein is cleaved using specific enzymes. The ionized amino acids are triggered by a laser beam which travels to the detector through a flight tube. The ions with lighter mass will reach the detector faster due to Newton's second law (F = ma) and hence, will be detected first. After the spectrum is recorded, it is further analyzed and compared against a database of sequenced proteins. A detailed sequence of protein fragments can be determined by repeating the process with different enzymes for cleavage. As a result, the fragments become much smaller with the fragments overlapping each other establishing the order of the protein.

2.2 Experimental determination of protein structure

The basic prerequisite for understanding the function of a protein is the knowledge of the protein 3D structure. The experimental methods used in the study of tertiary structure include:

i. *Protein X-ray crystallography*: X-ray crystallography is presently the most sought-after technique for determination of biological macromolecule structures. In this method, the determination of protein structure is achieved by crystallization of the purified protein at high concentration and exposing the crystals to an X-ray beam. The resultant diffraction patterns, obtained from the diffraction spots, are then processed to get knowledge about the symmetry of the packaging of the crystal and the size of the repeating units forming the crystal. A map of the electron density is then calculated using the "structure features", which are determined from the intensities of the diffraction spots. The quality of the electron density map can be improved using various methods. This is done to get a definitive idea to build the molecular structure using the amino acid sequence. Finally, the structure that is obtained is further refined to fit the map more accurately and to assume a conformation which is thermodynamically more favorable. Protein crystallography is known to provide highly accurate protein structures by giving atomic resolution. However, this method is not always straightforward and may take a lot of time to complete, which is around 3–5 years [11].

ii. *Nuclear magnetic resonance (NMR spectroscopy)*: Another useful technique to determine the protein structure is the NMR spectroscopy. It is a primary quantitative method which allows concentration determination of proteins in an aqueous environment that may resemble its actual physiological state more closely. In principle, the NMR spectroscopy is dependent on the electromagnetic radiation and the sample protein interaction. It is used to observe the local magnetic fields prevailing around the protein atomic nuclei. The NMR signal is obtained when sensitive radio receivers detect the excitation of the material nuclei with radio waves into the nuclear magnetic resonance. Thus, it provides access to the electronic structure of the sample protein. The major advantage of NMR over X-ray crystallography is that the protein in NMR spectroscopy can be examined in their native-like physiological state. However, NMR is not suitable for proteins with more than 150 amino acid and needs the protein under study to be stable in room temperature for a long time of data acquisition, which is a drawback of this technique [12].

iii. *Electron microscopy (especially Cryo-electron microscopy)*: Electron microscopy (EM) and cryo-electron microscopy (cryo-EM) are used to study objects that are comparatively larger in size such as cellular organelles or large macromolecular complexes with higher resolution. EM and cryo-EM use a method known as single-particle reconstruction. In principle, the data set in EM and cryo-EM is split randomly into half and the two averages (or 3D reconstructions) over rings (or shells, respectively) are compared, with increasing radius in Fourier space using an appropriate amount of reproducibility [13]. The protein sample in EM and cryo-EM does not require crystallization, saving a lot of time and effort, which is a major advantage over protein x-ray crystallography. Nevertheless, for membrane proteins, electron crystallography is used which require two dimensional (2D) crystals of the sample protein. Another advantage of cryo-EM is that it requires very less amount of sample materials. However, one of the limitations of cryo-EM is that it has to compromise with the resolution comparative to resolution obtained from x-ray crystallography and NMR spectroscopy [14].

2.3 Protein structure prediction

The field of structural biology is mostly dominated by experimental methods which are expensive and laborious in nature. However, since the last few decades, the application of computational techniques in structural biology has been widely used, with significant improvements in these techniques since last 10–20 years. This has helped to achieve substantial developments in protein structure prediction methods. *In-silico* protein structure prediction enables the prediction of 3D structures for proteins with known sequences and unknown structures. Prediction of the tertiary structure also helps in understanding the folding and unfolding of proteins. Also, protein engineering may help in incorporation of new functions in proteins thus facilitating drug design and discovery [15]. Protein structure prediction can be achieved by three different ways:

i. Computer simulation-based on empirical energy minimization

ii. Knowledge based-approaches using information derived from known sequences of experimentally determined protein 3-D structures

iii. Hierarchical methods.

2.3.1 Approaches based on energy minimization

The energy minimization method is also known as the *ab-initio* (*de* novo) method for protein structure prediction and is based on the theory that the native structure of protein is always at thermodynamic equilibrium with minimum energy, which is calculated using basic laws of physics and chemistry (**Figure 1**). Energy minimization-based methods always attempt to detect the global minima in free energy surface of the protein molecule as it is thought that global minima correspond to the native conformation. This method is not very helpful to design protein sequence length of more than 150 amino acid residues. However, it can be used to design small stable peptides that can bind to any specific therapeutic targets [16]. Two types of energy minimization methods are broadly used in *de novo* structure prediction approach, namely static and dynamical minimization methods. Some of the major FF used for energy minimizations are GROMOS, AMBER, CHARMM and ECEPP [17, 18]. One of the *ab-initio* protein structure prediction software packages is ROSETTA. This software package is based on the postulation that local interactions lead the conformation of short segments while global interactions establish the 3D protein structure [19]. The advantage of *ab-initio* approach is that it is based on physicochemical principles, however, these principles are hampered by the vast number of degrees of freedom which are needed to be looked after and also the performance of energy functions are

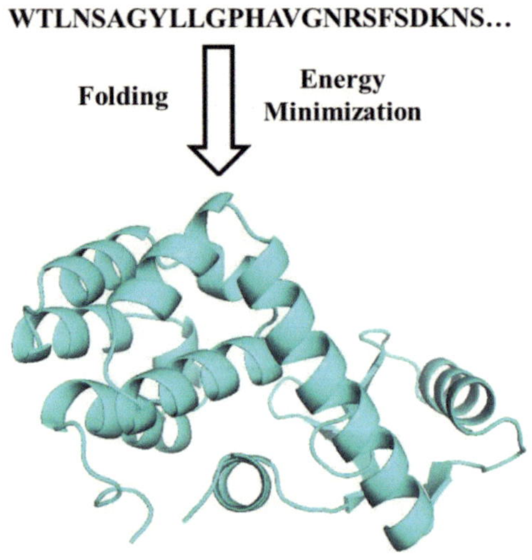

Figure 1.
An example of ab-initio structure prediction.

limited. The disadvantage of this method is that it requires high computations and for such studies there are no "good enough" interaction potentials which can model the native structure of a protein with atomic detail [20].

2.3.2 Approaches based on knowledge

The available protein structures are used to derive the knowledge based potentials [21, 22]. Further, these potentials are used to obtain the secondary structural information from amino acid sequence. The methods, based on the knowledge procured from known protein structures are of two types.

2.3.2.1 Homology modeling

One of the most powerful methods used to predict the 3D structure of proteins is the homology modeling. This method, also known as comparative modeling, uses a query protein having sequence similar with the target protein, having known tertiary structure [23–25]. The basis of this method lies on the observation that structures are more conserved than their sequences. Thus, if a target sequence has some degree of similarity with a protein sequence having known 3D structure, then that structure can be used to precisely model the target protein. A plethora of review articles are available on the strategies and challenges of computational protein structure prediction [8, 26].

For an accurate model building of a protein using homology modeling approach, the first step is template selection. The most crucial step involves the generation of a structure-based alignment between the query and the template protein sequence [27]. Models cannot be constructed for alignments having less than 20% identity. Additionally, the environment of the template such as the type of solvent, pH, presence of ligands, etc. and the quality of the experimentally-derived template structure must be taken into account. Once a desired template structure has been selected, a target-template alignment must be performed using standard sequence alignment techniques. After the creation of the template-target alignment, the 3D model of the target protein is created using several algorithms. Distance geometry is one of the commonly used methods to satisfy the spatial restraints obtained from the target-template alignment. MODELLER is one of the reliable homology modeling program and it imposes spatial restraints that are derived from the bond distances and angles in the target structure that are based on its alignment with the template structure, and stereo-chemical restraints on bond distance and dihedral angle preferences that are obtained from a representative set of all known protein structures. Then the constructed model is getting minimized using molecular dynamics to follow the spatial restraints [28].

After the creation of 3D model, the next step is to perform the quality assessment of the predicted model. From last few decades, many methods have been developed to assess the quality and correctness of modeled protein structures which analyze their stereochemistry. Some of the programs for such analysis are PROCHECK [29] and WhatCheck [30]. Another method to analyze the modeled protein is to calculate a residue-by-residue energy profile, where a peak in the profile corresponds to an error in the model. But this method has a drawback considering that a section of residues may appear to be inaccurate, while in reality they will be interacting with an incorrectly modeled region. Thus, for the assessment of modeled proteins, energy profile should not be the only means of identifying a good model.

Homology modeling for the prediction of protein 3D structures consists of multiple steps (**Figure 2**). Although a number of tools and web-servers are available, but no single server or tool can be considered as best in every aspect as compared to others. The function of a protein is dependent on the 3D structure; therefore, it is very important to enhance the quality of the predicted model. Homology modeling has a wide variety of applications in structural biology and plays a vital role in drug discovery process, as because for the study of drug-receptor (protein) interaction, the structure of the receptor (protein) is of utmost importance. However, this approach does not work if homologous structures are not available.

2.3.2.2 Threading

Threading, also known as fold recognition is a method that searches the protein structure template in a library of folds with the lowest possible energy for a given query sequence [15]. Fold recognition of a sequence requires a precise alignment of the query sequence corresponding to the positions of the amino acid residues of a folding motif. A set of possible positions of the amino acids in 3D space is

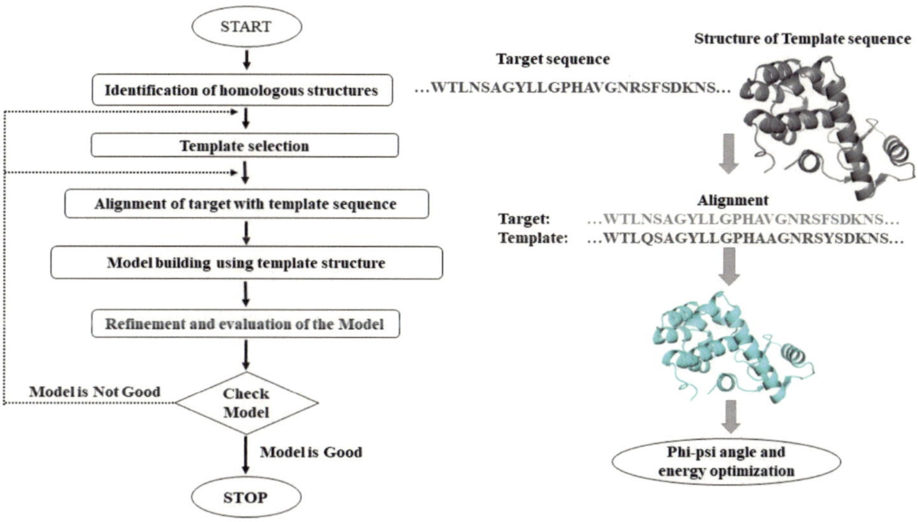

Figure 2.
A scheme of homology modeling.

established by the known structure. This step is followed by making a similar structure by placing the amino acids of the query sequence into their aligned positions. The main goal of this method is either to choose the most probable fold for any given sequence or to find out the appropriate sequences that have the possibility to fold into a given structure. This method is heavily dependent on the knowledge of experimental atomic details of the recognized protein folds and is generally applicable for only those proteins whose amino acid sequences adopt one of the protein folds that have already been experimentally established.

2.3.3 Approaches based on hierarchy

The Hierarchical approach is another strategy for protein structure prediction from their sequences. In principle, this method uses the hierarchy of protein structure, i.e., from the primary to secondary structure and secondary to tertiary structure. Thus, in order to understand the relationship of the primary amino acid sequence and the tertiary 3D structure, the intermediate secondary structure is predicted. This intermediate structure is used to build the tertiary 3D structure. A number of algorithms are developed for the modeling of secondary structure, but, unfortunately, the precision for prediction of secondary structures from their sequences is only about 80%. Currently the methods that are available for the secondary structure modeling can be divided into methods based on statistics, physicochemical properties, evolutionary information, combinatorial analysis and artificial intelligence [31–33].

2.4 Structure prediction methods and benchmarking

The performance assessment of existing methods is one of the major setbacks in the field of protein structure prediction as methods have been and are still in the process of development using different proteins with various evaluation criterions. Thus, in 1994, an open experiment was conducted all over the world with the intention of helping the developers and users of these methods. The experiment was called the Critical Assessment of Protein Structure Prediction (CASP) (https://predictioncenter.org/) [34]. The CASP is a community-wide, worldwide experiment which is conducted every two years since 1994. CASP allows research groups to test their structure prediction algorithms and establish the current state of the art in protein structure prediction. They help to identify the current progress as well as highlight the efforts that are needed to be addressed in the future.

3. Proteins: structure and function

Proteins are simple polymers of amino acids. The short stretches of polymers join together and get folded to form secondary structures which in turn give rise to the 3D structure of proteins. The secondary structures can be recognized either by hydrogen-bonding (H-bond) patterns among the carbonyl and amide groups in a peptide backbone or from the dihedral angles viz. phi and psi. Mainly two known secondary structures in a protein are α-helices and β-sheets which tend to build up into small repeating arrangements in protein structures; termed as 'supersecondary structures' or 'motifs'. These secondary structures assemble into larger subunits of structures termed as 'domains'. Domains can be further understood as the smallest structural unit of proteins which can be folded autonomously such as serine protease which is made up of two β barrel domains. Proteins comprises either of a single domain or multiple domains. Protein structures were for the first time categorized into folds in 1976 [35]. Murzein et al. later incorporated the idea and developed the publicly accessible database named SCOP (Structural Classification of Proteins) [36]. Folds in the SCOP were categorized by the class of secondary structure: all α, all β, α/β (wherein helices and sheets are mixed) and α + β (separate helices and sheets). Proteins are the most ubiquitous biomolecules and they accomplish the vast majority of functions in all the biological domains. The sequence-structure-function paradigm attracted the interests of scientists all over the world. As the proper functioning of all the biological processes depends on proteins and their non-functioning leads to grave diseases and disorder, biologists started working on them ever since. Way back in 1970s, Anfinsen have proposed that the 3D structure of native proteins comes from its sequence in a specified environment [37].

As proteins are dynamic in nature, experimental techniques fail to capture their different dynamical conformations and specially the transition between these conformations. One of the most widely utilized computational techniques, Molecular Dynamics (MD) Simulation tackles this challenge efficiently.

3.1 Molecular dynamics: the computational microscope

MD simulations assist us to comprehend and witness the time dependent behavior of proteins. As MD simulations have the ability to show the dynamic behavior of proteins at the level of atoms, it is also considered as computational microscope [38]. In this technique one requires an initial protein model which is obtained by either experimental methods or predictive modeling. As life sustains itself in water therefore one mimics simulation in explicit solvent. When the forces acting on all the atoms were acquired, Newton's laws of motion were utilized to compute the velocities and accelerations; besides updating the atom's positions. A time step of 2 fs (femtosecond) is usually applied for atomistic simulations while integrating the movement numerically. Finally, a trajectory of the system is generated by MD engine which can be further analyzed based on set objectives. The technique was first utilized in early 70's to study the most relevant biological challenge of the time; protein folding [39, 40]. The subsequent decades saw the application of MD simulations for investigating folding and unfolding mechanism of proteins [41]. Duan and Kollman were successful in 1998 to perform 1 μs MD simulation for the first time on parallel supercomputer. They investigated the protein folding mechanism of villin with explicit solvation [42]. Apart from proteins, the technique has been extended to study other relevant biomolecules [43, 44] and protein-nanoparticle interactions [45–49].

Simulation of any system revolves around lot of factors. Earlier the system size comprises of few thousand of atoms. With the advancement of both experimental and computational techniques, availability of 3D data in regard to proteins, proteins complexes, membrane proteins etc. has been possible which made the system size amplified to several lakhs of atoms with explicit solvent in consideration [50]. Meanwhile the advent of high-performance computing (HPC) and algorithm parallelization made it possible to run long timescale simulations for the above-mentioned systems. Further advancements in the algorithms of MD engines and/or the implementation of GPUs (graphical processing units) along with CPUs have significantly improved the performance of MD simulations. Some of the most popular simulation engines are: AMBER, CHARMM, DESMOND, GROMACS and NAMD. They have been integrated with messaging passing interface (MPI), which made it possible to utilize all the available cores of the computer simultaneously during a MD run to reduce the computation time.

3.2 Workhorse of simulation: the force fields

Force fields (FF) lie at heart of the MD simulation. In order to perform simulation, one needs the parameters to deduce the potential energy function [51].

The FF is a group of equations and associated parameters designed to imitate molecular geometry and selected properties of some tested molecules. FF comprises primarily of two components; bonded and non-bonded terms. Any molecular feature can be basically represented with them. The bonded terms can be represented by springs for bond length and angles along with torsional angles; the non-bonded terms comprise of Lennard-Jones potentials for van der Waals (vdW) interactions and Coulomb's law for electrostatic interactions. They were primarily developed to reproduce structural properties and applied to predict other properties such as thermodynamic parameters. Further the energy functions utilized in molecular mechanics commonly comprise topological parameters which are obtained from experiments or quantum mechanical calculations. An important feature of FF is transferability of the parameters and the functional form. It means to model a series of related molecules; the same set of parameters can be utilized rather than defining a new set of parameters for each individual molecule. Even though most of the FF are additive, a number of them having higher order terms are called class II FF. Some of widely utilized FF for bio-molecular simulations are AMBER, CHARMM, GROMOS and OPLS [52]. Additionally it is noteworthy to mention the application of FF in predicting structures of proteins/RNA. FFs were developed and benchmarked against experimentally solved structures and these FF were later incorporated to predict the structure for the ones lacking experimental information. Another important aspect of the FF is to discriminate the near-native protein conformation among the generated 3D models [53]. FFs are subject to rigorous scrutinizing and they were refined to improve their accuracy over time. One such example is the improvement of the residue side-chain torsion potentials of the Amber ff99SB FF which is also validated with available NMR experimental datasets [54]. A number of benchmark studies were conducted time to time, to compare different FFs. One difference arises among the available variety of FF is the bias/overestimate towards particular secondary structure of proteins. Man et al. recently concluded from their comparative simulation study that FFs (AMBER94, AMBER99 & AMBER12SB) were not able to predict β-sheet formation whereas FFs (AMBER96, GROMOS45a3, GROMOS53a5, GROMOS53a6, GROMOS43a1, GROMOS43a2, and GROMOS54a7) were able to form β-sheets swiftly. Further they have showed that the best FFs for investigating amyloid peptide assembly based on their structure and kinetics were AMBER99-ILDN, AMBER14SB, CHARMM22*, CHARMM36, and CHARMM36m [55].

3.3 Application examples of MD simulations

MD simulations have immensely contributed to solve and hypothesize many biological research problems. The significance of the computational microscope can be well understood by observing the increase in the vast repertoire of literature in the recent decade. The technique of simulation along with other computational tools plays a significant role in the field of protein structure prediction. Using a set of seven small proteins Kato et al. have validated the application of MD simulations to predict the 3D structure of proteins. The set of small proteins were in the range of 10 to 46 residues. They have considered two properties; root mean squared deviation (RMSD) and occurrence of secondary structure to validate the predicted structures from simulation with that of the available experimental ones. AMBER12 simulation package with AMBER ff12SB have been utilized to carry out their simulations. With the help of MD simulations, they have shown the possibility of reproducing the secondary structures of small proteins [56]. Our group has also utilized the indispensable technique of simulation recently to investigate the dynamics and stability of *ab-initio* predicted structure of bacterial effector protein, HopS2. The importance of the effector proteins lies with them conferring pathogenicity to bacteria. As the sequence similarity of the effector proteins lies in the twilight zone along with the few partially solved structures of effector proteins at disposal, it is a perplexing task to study the sequence-structure-function relationship of these proteins. With the assistance of MD simulations, our group was able to show the stability of local secondary structural elements of HopS2 which are vital to its overall structure and interaction. These investigation have been performed using Gromacs along with OPLS FF [57].

Another interesting aspect of human proteome is the intrinsically disordered proteins (IDP). There are many examples of proteins with folded domains but they feature disordered regions while some are entirely unstructured. Some IDPs fold upon interacting with their binding partners while other persists in unfolded state even in a bound complex. The IDPs plays a critical role in cell signaling and regulation. Pietrek et al. have carried out a recent work in this direction. They have considered a hierarchical algorithm to generate large ensembles of full length IDP structures and these structures can be further used as starting points for atomistic simulations. The IDP structures generated by their hierarchical approach implemented with all atom MD simulations were able to capture both local conformations compared with NMR experiments and also the gross dimension described by small angle X-ray experiments. Gromacs simulation package along with Amber03ws and Amber99SB*-ILDN-q FF were utilized by them to carry out the investigation [58]. The powerful computational microscope was also applied to investigate structure and dynamics of plasma membrane proteins. Mattedi et al. recently utilized MD simulations to study glucagon receptor, a class B GPCR. The glucagon-induced release of glucose from the liver into the bloodstream is facilitated by the glucagon receptor. There is scarce information about the mechanism of this receptor. They utilized extensive MD simulations and free energy landscape computation to elucidate the activation mechanism of the receptor. Through their simulation work, they identified an intermediate state of the glucagon receptor and decipher the mechanism of allosteric antagonists of the glucagon which locks transmemebrane helix 6. They have employed AMBER14SB FF and LipidBook parameters for lipids with Gromacs package in their work [59].

4. Molecular docking

The plethora of diseases discovered ever since and being investigated tirelessly by scientists all over the world ultimately culminates to the sole objective of finding effective solutions. The therapeutic targets in most of the cases are proteins. After knowing their mechanism of actions, how the proteins works and what goes wrong during the diseased state, the next notion is to challenge their functionality with designing some inhibitors. It comes under the domain of drug discovery. And one of the most challenging fields of study is the drug design and development. The complete clinical trials take about 10–15 years of time with billions of dollars expenses for a single drug to reach market. With the completion of human genome project which leads to identification of ever-increasing number of new drug targets (mainly proteins); the efforts were strengthened to find solution to the diseases. Additionally, the availability of 3D structures of protein and protein-ligand complexes made it feasible to carry out research in this area. However, to experimentally screen millions of compounds and their conformers for a single therapeutic target requires enormous amount of time and resources which makes it quite challenging. With the application of computational techniques, the pre-clinical period can be reduced to save valuable assets. The *in-silico* approaches will significantly curtail the time needed for hit identification and also improve the chances of finding the anticipated drug molecules. To facilitate drug design and discovery, several modeling techniques were available and mostly they are categorized into two main approaches viz. structure-based and ligand-based drug design approaches. The structure-based approach mainly relies on the 3D data of target and the ligand. The ligand-based approach is chiefly adopted in the absence of known experimental structure of the target. In ligand-based approach, the known ligands which were bound to the targets were investigated to decipher the physiochemical and structural properties of the ligands and these were correlated with the anticipated pharmacological activity of the ligands in hand [60].

One of the most extensively utilized computational techniques in the structure-based drug design is molecular docking. Molecular docking is usually achieved by first predicting the molecular orientation or pose of a ligand within the active site of a target and followed by assessing their binding affinity with the usage of a scoring function. The technique is exploited to decipher the interactions between a target and ligand at the atomic level allowing us to describe the behavior of ligands within the active sites of targets as well as to reveal fundamental biochemical processes. Since the first developments of docking algorithms in the 1980s, molecular docking became an indispensable tool in the field of drug discovery [61].

4.1 Types of molecular docking

Molecular docking can be basically categorized into three types: rigid docking, semi-flexible docking and flexible docking. In the rigid docking approach, both the structure of target and ligand does not change. The computation method is relatively modest and chiefly spans the degree of conformational matching, thus it is more apt for investigating macromolecular systems such as protein-nucleic acid and protein-protein systems. The semi/quasi flexible docking approach take flexibility into consideration while docking of the ligand and thus it is more appropriate to deal with the intermolecular interactions of small molecules and proteins. Usually the structure of the ligands can move freely while the target remains rigid or retain few rotatable residues ensuring computational efficiency during the docking process. In the flexible docking method, it is based on the idea that a protein is not always a rigid entity during the course of ligand binding and thus it considers both the protein and ligand as flexible entities. Over the years various methods have been introduced, based on induced fit model and/or conformational sampling.

4.1.1 Scoring function

One crucial element of any docking algorithm is the scoring function. The scoring function aids in the pose selection and it is involved in distinguishing putative precise binding modes and to filter out the non-binders from the *N* number of generated poses during a docking run. The speed and accuracy of docking programs is also dependent on scoring functions. Further computational efficiency and reliability are points kept in mind while developing any scoring function. There are three categories of scoring functions:

i. Force-field based scoring function

This scoring function is based on the concept of molecular mechanics which estimates the potential energy of a system with a mixture of intramolecular and intermolecular elements. In molecular docking, the intermolecular elements are usually considered, with the probable ligand-bonded terms, especially the torsional constituents. The non-bonded constituents include the van der Waals term which is defined by Lennard-Jones potential, and the electrostatic term, specified by the Coulomb function. GoldScore [62], AutoDock [63] and GBVI/WSA [60] are few examples of the mentioned scoring function.

ii. Empirical scoring function

Empirical function is the sum of different empirical terms such as van der Waals, H-bond, electrostatic, entropy, desolvation, hydrophobicity, etc. Utilizing least square fitting method, they are optimized on a training set of target-ligand complexes to reproduce the binding affinity data. Empirical scoring functions compared to force-field ones are computationally much more efficient owing to their simple energy terms. The first example of empirical scoring function is the LUDI scoring function [64]. GlideScore [65] and ChemScore [66] are other examples of empirical scoring functions.

iii. Knowledge-based scoring function

Knowledge-based functions are directly obtained from the structural information of experimentally solved protein-ligand complexes. The frequencies of interatomic contact and/or distances between the target and the ligand are obtained. The premise for this criterion relies on the assumption that frequency of occurrences will be greater for the ones with more favorable interactions. Pairwise atom-type potentials were generated with the obtained frequency distributions. Further the score is computed by preferred interactions and imposing penalty for repulsive contacts between each pair of atoms in the target and ligand within a set cutoff. Examples of this scoring function are DrugScore [67] and GOLD/ASP functions [68].

With the advancement in the field of high-performance computing, scientists have also applied artificial intelligence based and machine learning based scoring functions in virtual screening which holds promising outcomes [69].

4.1.2 Sampling algorithms

Sampling plays the next crucial role in any molecular docking program. With a set therapeutic target, the sampling algorithm will generate a number of conformations (poses) of the small molecule within the docked site of the target. The knowledge of the docked site is considered either from experimental data or predicted with the aid of active site prediction software. As the speed and accuracy of molecular docking plays a role in large virtual screening research works, the area of developing and/or improving existing sampling algorithms have provided ample opportunities for computational scientists. The sampling algorithms can be categorized as: shape matching, systematic search algorithm and stochastic algorithm.

i. Shape matching

One of the earliest methods designed was the shape matching algorithm for sampling. The criterion implemented in this algorithm is that the molecular surface of the small molecule needs to complement the molecular surface of the binding region of the target. The three translational and three rotational (six degree of freedom) of the small molecule led to spans many probable orientations. Thus, the goal of this algorithm is to place as smoothly and quickly the small molecule into the binding site based on shape complementarity. In this method, the conformation of the small molecule is usually fixed and therefore, this method along with flexible-docking is usually preferred rather than only shape matching. DOCK [61], LigandFit [70] and Surflex [70] are few examples of docking programs where shape matching algorithm is used.

ii. Systematic Search

With the help of systematic search algorithm, the ligand can explore all the degrees of freedom and it can generate all probable conformations. Unlike in shape matching algorithm, the conformations of ligands are not fixed here. Systematic search technique can be categorized into three types: exhaustive search, fragmentation and conformational ensemble.

In exhaustive search method, all the rotatable bonds of the small molecules are scanned in a systematic manner. However, to avoid a huge combinatorial explosion & to make the docking procedure practical, the search space is limited by geometric constraints criterion. Glide docking program implements this method.

The fragmentation method as the name suggests implements the idea of fragmenting the ligands into smaller rigid fragments. The incremental construction is one such mode wherein one fragment is placed first in the binding site and other fragments were attached incrementally. FlexX [70] utilizes this algorithm.

In the conformational ensemble algorithm, small molecule flexibility is signified by rigidly docking an ensemble of pre-generated conformers of the small molecule. Next the binding modes were collected from different docking runs then binding energy values are used to rank them. FLOG [71] and MS-DOCK [70] implements this algorithm.

iii. Stochastic Search

In the stochastic search, the sampling of the small molecule conformations is carried out by making random changes at every step in both the rotational/translation space and conformational space of the small molecule respectively. A probabilistic criterion is placed to either accept or reject the random change. Within stochastic search, there are four subtypes viz., Monte Carlo method, evolutionary algorithms (EA), Tabu search methods and swarm optimization (SO) methods. Genetic algorithm, one type of EA is implemented in AutoDock [63] and GOLD docking programs.

It is imperative to mention here that different docking programs/servers apply variety of algorithms in multi-phase wise in their docking pipeline.

4.2 Application examples of molecular docking

The molecular docking can be seen applied regularly in academic labs and pharmaceutical companies to find effective solutions and thwart deadly diseases [72]. The identification of hit molecules in the preliminary stage of drug discovery is today heavily relied upon high throughput screening. Moreover, the availability of small molecule databases such as PubChem, ZINC, MayBridge etc. along with the growth of experimental structures of targets (proteins, membrane proteins, protein-ligand complexes) have made the use of molecular docking to screen millions of compounds and made it possible to test only lead molecules.

G protein-coupled receptors (GPCR) are the attractive targets of drug design regimes because of their importance in cell signaling and functions. Kolb et al. have considered β2-adrenergic receptor, a GPCR found in the smooth muscle tissue to investigate the structure-based approach for ligand discovery. In their study, they have utilized DOCK molecular docking program to screen approx. 1 million compounds from ZINC database. They were able to test experimentally the resultant 25 high ranked molecules from docking; of which 6 molecules showed binding affinity <4 μM. And the best compound showed 9 nM of inhibition constant against the receptor [73].

Rajkhowa et al. have utilized the structure-based drug design (SBDD) method along with MD simulations to design inhibitors against malaria, one of the most devastating infectious diseases. They have considered 178 compounds similar to known anti-malarial imidazopyrazine from the PubChem database to carry out the work. The target of the inhibitor is the phosphatidylinositol-4-OH kinase which is a lipid kinase involved in the membrane ingestion process of the erythrocytic stage of the life cycle of the plasmodium and recognized as a drug target. AutoDock 4.2 has been utilized in their work. They have reported three potential inhibitors based on molecular docking, MD simulations and ADMET studies [74].

Our group had worked in the direction of SBDD to tackle insulin resistance and type-2 diabetes (T2D). We have considered 142 anti-diabetic compounds spanning various categories of phytochemicals such as flavonoids, alkanoids, sulfonylurea and terpenes. The target of the study is A_{2A} adenosine receptor which had been shown in reports that it can be utilized to counteract insulin resistance and adipocyte inflammation. Numerous computational tools were utilized to carry out the work such as druglikeness filtering, QSAR modeling, ADMET profiling to molecular docking. The different level of screenings led to 6 molecules which were docked with the help of two different molecular docking approaches viz. AutoDock and AutoDockFR to get optimal receptor-ligand conformations. From the 142 compounds finally we got one molecule “indirubin-3′-monoxime” which is then followed by experimental validations [75].

5. Conclusion

In this era of high-performance computing technology, there is hardly any field of science which is not touched upon by some amount of significant computational works. The potential of computing power is much reliant on advancement in hardware and algorithms. Substantial number of computational tools and techniques were developed and applied in the fascinating area of proteomics also. Mathematical models were devised in the form of FF parameters and implemented in various algorithms. Here, we have discussed the inevitable role of FF in protein structure prediction/modeling, conformational dynamics and their functional aspects along with the applications in virtual screening programs. As discussed in the chapter, a lot of programs with variety of FFs are available for structure prediction, MD simulations etc., but there is still a scope of further developments. For example, till now it is a challenge for accurately predicting protein structures of larger sizes or the protein sequences having low amount of similarity with sequences of known structures. Also, the existing software are in use for trans-membrane protein structure prediction but it is an hour need to develop different program to model the trans-membrane segments. Although MD simulations were utilized for validating predicted structures of membrane proteins and/or for getting insights of their mechanism, challenge remains in the forms of FF as at times it is difficult to get the parameters for membrane proteins, lipids in which they were embedded, any bound coordinated metal ions in a single FF. The accuracy of models depends upon pH and dynamic charge environment instead of static electrostatic charges, and polarizable water models, requires further development and testing of polarizable force fields. The existing FF were designed with aid of experimental data for globular proteins and applied for studying IDPs whereas disordered proteins are having non-structural segments. Thus, it necessitates designing and developing different set of FF parameters for simulating exclusively IDPs. In summary, there is always space for improvement in existing ones and developing new models with higher accuracy in any field of science.

Acknowledgements

Authors would like to acknowledge Department of Biotechnology (DBT) (project number BT/COE/34/SP28408/2018), Govt. of India for providing computational facilities.

Author details

Zaved Hazarika[1], Sanchaita Rajkhowa[2] and Anupam Nath Jha[1*]

1 Department of Molecular Biology and Biotechnology, Tezpur University, Tezpur, Assam, India

2 Centre for Biotechnology and Bioinformatics, Dibrugarh University, Dibrugarh, Assam, India

*Address all correspondence to: anjha@tezu.ernet.in

References

[1] Nii-Trebi NI. Emerging and Neglected Infectious Diseases: Insights, Advances, and Challenges. Biomed Res Int. 2017;2017:5245021.

[2] Genetics Home Reference Available from: https://ghr.nlm.nih.gov/condition/huntington-disease#genes. Accessed on 2020-07-16.

[3] World Health Organization, "Infections and infectious diseases: a manual for nurses and midwives in the WHO European Region," Available from: http://www.euro.who.int/__data/assets/pdf_file/0013/102316/e79822.pdf. Accessed on 2020-07-16.

[4] Understanding Emerging and Re-emerging Infectious Diseases. National Institutes of Health (US): National Institutes of Health (US); Biological Sciences Curriculum Study. NIH Curriculum Supplement Series 2007.

[5] Gorbalenya AE, Baker SC, Baric RS, de Groot RJ, Drosten C, Gulyaeva AA, et al. The species Severe acute respiratory syndrome-related coronavirus: classifying 2019-nCoV and naming it SARS-CoV-2. Nature Microbiology. 2020;5(4):536-44.

[6] Lu H, Stratton CW, Tang YW. Outbreak of pneumonia of unknown etiology in Wuhan, China: The mystery and the miracle. J Med Virol. 2020;92(4):401-2.

[7] Zhu N, Zhang D, Wang W, Li X, Yang B, Song J, et al. A Novel Coronavirus from Patients with Pneumonia in China, 2019. N Engl J Med. 2020;382(8):727-33.

[8] Deng H, Jia Y, Zhang Y. Protein structure prediction. Int J Mod Phys B. 2018;32(18).

[9] Berman HM, Vallat B, Lawson CL. The data universe of structural biology. IUCrJ. 2020;7(4):630-8.

[10] Berg JM, Tymoczko JL, Stryer L. Biochemistry, 7th ed: W.H. Freeman and Company; 2012.

[11] Smyth MS, Martin JH. x ray crystallography. Mol Pathol. 2000;53(1):8-14.

[12] Marion D. An introduction to biological NMR spectroscopy. Molecular & cellular proteomics : MCP. 2013;12(11):3006-25.

[13] Frank J. Single-particle reconstruction of biological macromolecules in electron microscopy--30 years. Q Rev Biophys. 2009;42(3):139-58.

[14] Milne JL, Borgnia MJ, Bartesaghi A, Tran EE, Earl LA, Schauder DM, et al. Cryo-electron microscopy--a primer for the non-microscopist. FEBS J. 2013;280(1):28-45.

[15] Kumar P, Halder S, Bansal M. Biomolecular Structures: Prediction, Identification and Analyses. In: Ranganathan S, Gribskov M, Nakai K, Schönbach C, editors. Encyclopedia of Bioinformatics and Computational Biology. Oxford: Academic Press; 2019. p. 504-34.

[16] Chevalier A, Silva DA, Rocklin GJ, Hicks DR, Vergara R, Murapa P, et al. Massively parallel de novo protein design for targeted therapeutics. Nature. 2017;550(7674):74-9.

[17] Pearlman DA, Case DA, Caldwell JW, Ross WS, Cheatham TE, DeBolt S, et al. AMBER, a package of computer programs for applying molecular mechanics, normal mode analysis, molecular dynamics and free energy calculations to simulate the structural and energetic properties of molecules. Computer Physics Communications. 1995;91(1):1-41.

[18] van Gunsteren WF, Berendsen HJC. Computer Simulation of Molecular Dynamics: Methodology, Applications, and Perspectives in Chemistry. Angewandte Chemie International Edition in English. 1990;29(9):992-1023.

[19] Simons KT, Bonneau R, Ruczinski I, Baker D. Ab initio protein structure prediction of CASP III targets using ROSETTA. Proteins. 1999;Suppl 3:171-6.

[20] Bonneau R, Baker D. Ab initio protein structure prediction: progress and prospects. Annu Rev Biophys Biomol Struct. 2001;30:173-89.

[21] Jha AN, Vishveshwara S, Banavar JR. Amino acid interaction preferences in proteins. Protein Sci. 2010;19(3):603-16.

[22] Jha AN, Vishveshwara S, Banavar JR. Amino acid interaction preferences in helical membrane proteins. Protein Eng Des Sel. 2011;24(8):579-88.

[23] Blundell TL, Sibanda BL, Sternberg MJ, Thornton JM. Knowledge-based prediction of protein structures and the design of novel molecules. Nature. 1987;326(6111):347-52.

[24] Šali A, Overington JP, Johnson MS, Blundell TL. From comparisons of protein sequences and structures to protein modelling and design. Trends in Biochemical Sciences. 1990;15(6):235-40.

[25] Sutcliffe MJ, Haneef I, Carney D, Blundell TL. Knowledge based modelling of homologous proteins, part I: three-dimensional frameworks derived from the simultaneous superposition of multiple structures. Protein Engineering, Design and Selection. 1987;1(5):377-84.

[26] Kc DB. Recent advances in sequence-based protein structure prediction. Briefings in Bioinformatics. 2016;18(6):1021-32.

[27] Pascarella S, Argos P. A data bank merging related protein structures and sequences. Protein Eng. 1992;5(2):121-37.

[28] Farokhirad S, Bradley RP, Sarkar A, Shih A, Telesco S, Liu Y, et al. 3.13 Computational Methods Related to Molecular Structure and Reaction Chemistry of Biomaterials☆. In: Ducheyne P, editor. Comprehensive Biomaterials II. Oxford: Elsevier; 2017. p. 245-67.

[29] Laskowski RA, MacArthur MW, Moss DS, Thornton JM. PROCHECK: a program to check the stereochemical quality of protein structures. Journal of Applied Crystallography. 1993;26(2):283-91.

[30] Hooft RW, Vriend G, Sander C, Abola EE. Errors in protein structures. Nature. 1996;381(6580):272.

[31] Rost B. PHD: predicting one-dimensional protein structure by profile-based neural networks. Methods Enzymol. 1996;266:525-39.

[32] Cuff JA, Clamp ME, Siddiqui AS, Finlay M, Barton GJ. JPred: a consensus secondary structure prediction server. Bioinformatics. 1998;14(10):892-3.

[33] Kaur H, Raghava GP. Prediction of alpha-turns in proteins using PSI-BLAST profiles and secondary structure information. Proteins. 2004;55(1):83-90.

[34] Kryshtafovych A, Schwede T, Topf M, Fidelis K, Moult J. Critical assessment of methods of protein structure prediction (CASP)—Round XIII. Proteins: Structure, Function, and Bioinformatics. 2019;87(12):1011-20.

[35] Levitt M, Chothia C. Structural patterns in globular proteins. Nature. 1976;261(5561):552-8.

[36] Andreeva A, Howorth D, Chothia C, Kulesha E, Murzin AG. SCOP2 prototype: a new approach to protein structure mining. Nucleic Acids Res. 2014;42(Database issue):D310-4.

[37] Anfinsen CB. Principles that govern the folding of protein chains. Science. 1973;181(4096):223-30.

[38] Lee EH, Hsin J, Sotomayor M, Comellas G, Schulten K. Discovery through the computational microscope. Structure. 2009;17(10):1295-306.

[39] Levitt M, Warshel A. Computer simulation of protein folding. Nature. 1975;253(5494):694-8.

[40] McCammon JA, Gelin BR, Karplus M. Dynamics of folded proteins. Nature. 1977;267(5612):585-90.

[41] Karplus M, Sali A. Theoretical studies of protein folding and unfolding. Curr Opin Struct Biol. 1995;5(1):58-73.

[42] Duan Y, Kollman PA. Pathways to a protein folding intermediate observed in a 1-microsecond simulation in aqueous solution. Science. 1998;282(5389):740-4.

[43] Das S, Hazarika Z, Sarmah S, Baruah K, Rohman MA, Paul D, et al. Exploring the interaction of bioactive kaempferol with serum albumin, lysozyme and hemoglobin: A biophysical investigation using multi-spectroscopic, docking and molecular dynamics simulation studies. Journal of Photochemistry and Photobiology B: Biology. 2020;205:111825.

[44] Das S, Sarmah S, Hazarika Z, Rohman MA, Sarkhel P, Jha AN, et al. Targeting the heme protein hemoglobin by (−)-epigallocatechin gallate and the study of polyphenol–protein association using multi-spectroscopic and computational methods. Physical Chemistry Chemical Physics. 2020;22(4):2212-28.

[45] Hazarika Z, Jha AN. Computational Analysis of the Silver Nanoparticle–Human Serum Albumin Complex. ACS Omega. 2020;5(1):170-8.

[46] Nayak PS, Borah SM, Gogoi H, Asthana S, Bhatnagar R, Jha AN, et al. Lactoferrin adsorption onto silver nanoparticle interface: Implications of corona on protein conformation, nanoparticle cytotoxicity and the formulation adjuvanticity. Chemical Engineering Journal. 2019;361:470-84.

[47] Arakha M, Borah SM, Saleem M, Jha AN, Jha S. Interfacial assembly at silver nanoparticle enhances the antibacterial efficacy of nisin. Free Radical Biology and Medicine. 2016;101:434-45.

[48] Saikia N, Jha AN, Deka RC. Interaction of pyrazinamide drug functionalized carbon and boron nitride nanotubes with pncA protein: a molecular dynamics and density functional approach. Rsc Advances. 2013;3(35):15102-7.

[49] Asthana S, Hazarika Z, Nayak PS, Roy J, Jha AN, Mallick B, et al. Insulin adsorption onto zinc oxide nanoparticle mediates conformational rearrangement into amyloid-prone structure with enhanced cytotoxic propensity. Biochimica et Biophysica Acta (BBA) - General Subjects. 2019;1863(1):153-66.

[50] Bora N, Jha AN. An integrative approach using systems biology, mutational analysis with molecular dynamics simulation to challenge the functionality of a target protein. Chemical Biology & Drug Design. 2019;93(6):1050-60.

[51] Rajkhowa S, Jha AN. Molecular Dynamics Simulations: A Tool to Investigate the Interactions between Biomolecules and Nanoparticles. Nova Science Publishers. 2019:65-108.

[52] Fluitt Aaron M, de Pablo Juan J. An Analysis of Biomolecular Force Fields for Simulations of Polyglutamine in Solution. Biophysical Journal. 2015;109(5):1009-18.

[53] Rubenstein AB, Blacklock K, Nguyen H, Case DA, Khare SD. Systematic Comparison of Amber and Rosetta Energy Functions for Protein Structure Evaluation. Journal of chemical theory and computation. 2018;14(11):6015-25.

[54] Lindorff-Larsen K, Piana S, Palmo K, Maragakis P, Klepeis JL, Dror RO, et al. Improved side-chain torsion potentials for the Amber ff99SB protein force field. Proteins. 2010;78(8):1950-8.

[55] Man VH, He X, Derreumaux P, Ji B, Xie XQ, Nguyen PH, et al. Effects of All-Atom Molecular Mechanics Force Fields on Amyloid Peptide Assembly: The Case of Abeta16-22 Dimer. Journal of chemical theory and computation. 2019;15(2):1440-52.

[56] Kato K, Nakayoshi T, Fukuyoshi S, Kurimoto E, Oda A. Validation of Molecular Dynamics Simulations for Prediction of Three-Dimensional Structures of Small Proteins. Molecules. 2017;22(10).

[57] Borah SM, Jha AN. Identification and analysis of structurally critical fragments in HopS2. BMC Bioinformatics. 2019;19(Suppl 13):552.

[58] Pietrek LM, Stelzl LS, Hummer G. Hierarchical Ensembles of Intrinsically Disordered Proteins at Atomic Resolution in Molecular Dynamics Simulations. Journal of chemical theory and computation. 2020;16(1):725-37.

[59] Mattedi G, Acosta-Gutierrez S, Clark T, Gervasio FL. A combined activation mechanism for the glucagon receptor. Proc Natl Acad Sci U S A. 2020;117(27):15414-22.

[60] Corbeil CR, Williams CI, Labute P. Variability in docking success rates due to dataset preparation. J Comput Aided Mol Des. 2012;26(6):775-86.

[61] Kuntz ID, Blaney JM, Oatley SJ, Langridge R, Ferrin TE. A geometric approach to macromolecule-ligand interactions. J Mol Biol. 1982;161(2):269-88.

[62] Verdonk ML, Cole JC, Hartshorn MJ, Murray CW, Taylor RD. Improved protein–ligand docking using GOLD. *Proteins: Structure, Function, and Bioinformatics*. 2003;52(4):609-23.

[63] Morris GM, Goodsell DS, Halliday RS, Huey R, Hart WE, Belew RK, et al. Automated docking using a Lamarckian genetic algorithm and an empirical binding free energy function. Journal of Computational Chemistry. 1998;19(14):1639-62.

[64] Böhm H-J. The development of a simple empirical scoring function to estimate the binding constant for a protein-ligand complex of known three-dimensional structure. Journal of Computer-Aided Molecular Design. 1994;8(3):243-56.

[65] Friesner RA, Murphy RB, Repasky MP, Frye LL, Greenwood JR, Halgren TA, et al. Extra precision glide: docking and scoring incorporating a model of hydrophobic enclosure for protein-ligand complexes. *J Med Chem*. 2006;49(21):6177-96.

[66] Eldridge MD, Murray CW, Auton TR, Paolini GV, Mee RP. Empirical scoring functions: I. The development of a fast empirical scoring function to estimate the binding affinity of ligands in receptor complexes. J Comput Aided Mol Des. 1997;11(5):425-45.

[67] Velec HF, Gohlke H, Klebe G. DrugScore(CSD)-knowledge-based scoring function derived from small molecule crystal data with superior recognition rate of near-native ligand

poses and better affinity prediction. J Med Chem. 2005;48(20):6296-303.

[68] Mooij WT, Verdonk ML. General and targeted statistical potentials for protein-ligand interactions. Proteins. 2005;61(2):272-87.

[69] Sieg J, Flachsenberg F, Rarey M. In Need of Bias Control: Evaluating Chemical Data for Machine Learning in Structure-Based Virtual Screening. Journal of Chemical Information and Modeling. 2019;59(3):947-61.

[70] Pagadala NS, Syed K, Tuszynski J. Software for molecular docking: a review. Biophysical reviews. 2017;9(2):91-102.

[71] Miller MD, Kearsley SK, Underwood DJ, Sheridan RP. FLOG. a system to select 'quasi-flexible' ligands complementary to a receptor of known three-dimensional structure. J Comput Aided Mol Des. 1994;8(2):153-74.

[72] Rajkhowa S, Jha AN, Deka RC. Anti-tubercular drug development: computational strategies to identify potential compounds. J Mol Graph Model. 2015;62:56-68.

[73] Kolb P, Rosenbaum DM, Irwin JJ, Fung JJ, Kobilka BK, Shoichet BK. Structure-based discovery of beta2-adrenergic receptor ligands. Proc Natl Acad Sci U S A. 2009;106(16):6843-8.

[74] Rajkhowa S, Borah SM, Jha AN, Deka RC. Design of *Plasmodium falciparum* PI(4)KIIIβ Inhibitor using Molecular Dynamics and Molecular Docking Methods. ChemistrySelect. 2017;2(5):1783-92.

[75] Choudhary SA, Bora N, Banerjee D, Arora L, Das AS, Yadav R, et al. A novel small molecule A2A adenosine receptor agonist, indirubin-3'-monoxime, alleviates lipid-induced inflammation and insulin resistance in 3T3-L1 adipocytes. Biochem J. 2019;476(16):2371-91.

Section 3

Research Chapters

Chapter 4

Importance of Homology Modeling for Predicting the Structures of GPCRs

Ananthasri Sailapathi, Seshan Gunalan, Kanagasabai Somarathinam, Gugan Kothandan and Diwakar Kumar

Abstract

Homology modeling is one of the key discoveries that led to a rapid paradigm shift in the field of computational biology. Homology modeling obtains the three dimensional structure of a target protein based on the similarity between template and target sequences and this technique proves to be efficient when it comes to studying membrane proteins that are hard to crystallize like GPCR as it provides a higher degree of understanding of receptor-ligand interaction. We get profound insights on structurally unsolved, yet clinically important drug targeting proteins through single or multiple template modeling. The advantages of homology modeling studies are often used to overcome various problems in crystallizing GPCR proteins that are involved in major disease-related pathways, thus paving way to more structural insights via in silico models when there is a lack of experimentally solved structures. Owing to their pharmaceutical significance, structural analysis of various GPCR proteins using techniques like homology modeling is of utmost importance.

Keywords: membrane protein, bovine rhodopsin, template-based modeling, GPCR-EXP, GPCRdb

1. Introduction

The comparative modeling of proteins, more popularly known as homology modeling among the research community, is a computational procedure that constructs three dimensional atomic resolution structure of a 'target' protein, the structure of it is unknown. A new structure for the target protein is modeled using its own amino acid sequence and a known experimental structure of a homologous protein as a template based upon which the model is constructed. This template-based modeling technique became a plausible computational technique because of the fact that evolutionary related proteins share a similar structure [1]. This undeniable truth led to the famous outbreak of using homology modeling to determine the three dimensional structure of proteins whose structures were otherwise difficult to solve.

One such family of protein that poses a great challenge to study is membrane proteins due to their partially flexibility and lack of stability. The surface of membrane proteins is also comparatively hydrophobic and can only be extracted from the cell membrane with detergents which cause challenges at many levels, including expression, solubilization, purification, crystallization, data collection and structure solution. **Figure 1** shows the total number of membrane protein structures deposited in PDB as of August 2020 and this data was derived from the mpstruc database [2]. There are 2037 published reports of membrane protein structures in this database. It is also very clear that the number of available membrane protein structures is very less compared to the expected exponential growth in the number of available structures [3]. Though approximately 25% of all proteins are membrane proteins, there are less solved structures available due to the difficulty in crystallizing membrane proteins.

Many advances are being made in developing novel methods that can help in solving and studying the structure of membrane proteins in a high-throughput manner. The key to overcome the membrane protein structural biology is the underlying fact that they are structurally homologous to proteins which are evolutionarily related to them. This kindled the structural biologists to try a large number of targets and homologs of each target so that at least a few proteins will show progress through all the steps associated with their structural studies. This is where computational techniques like homology modeling came to the aid of structural biologists in helping solve the structures of membrane proteins by obtaining the three dimensional structure of a target protein based on the similarity between template and target sequences. One arena of such membrane protein structural biology research that has proved to be promising is the G-protein-coupled receptors (GPCRs) which are the largest family of membrane proteins.

The GPCRs constitute a diverse family of proteins in mammalian genomes [4]. The first GPCR for which structure was determined was Rhodopsin, a prototypical class A GPCR. The GPCRs are categorized into five major classes based on their

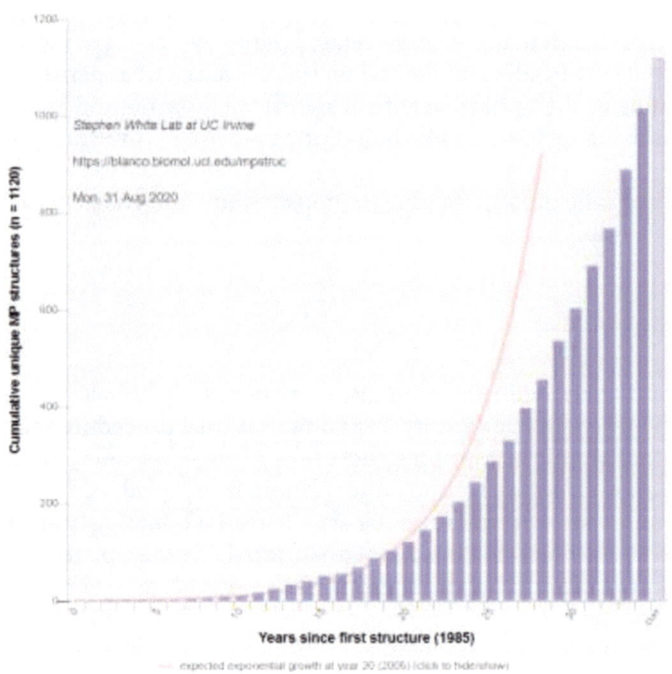

Figure 1.
The number of membrane protein structures deposited each year since the first structure was solved as given by the mpstruc database. The number of structures available is considerably low compared to the expected number of membrane protein structures (red line).

sequences as well as on their known or suspected functions in vertebrate: rhodopsin (family A), secretin (family B), glutamate (family C), adhesion and Frizzled/Taste2 [5]. The actual estimate of GPCRs in human genome is still being analyzed. The presence of seven transmembrane (7-TM) spanning α-helical segments separated by alternating intracellular and extracellular loop regions is one of the characteristic features in the structure of GPCRs. They also possess an extracellular N-terminus and an intracellular C-terminus which paved way for GPCRs to be also known as the 7-TM receptors or the heptahelical receptors. The tertiary structure of the GPCR resembles a barrel, with the seven transmembrane helices forming a cavity within the plasma membrane that serves as a ligand-binding domain. With its unique structure, the GPCRs serve many important roles in the human body. Hence the structure function correlation of GPCRs is a vital area of research even today.

The crystal structure of protein plays a pivot role in determining the functional importance of a protein. However membrane proteins are difficult to crystallize. Being a membrane protein, the GPCR structural studies have complexity because of low protein expression level in native tissues and heterologous systems. The poor protein stability and multiple conformational states of the receptors also are major hurdles in the GPCR structural studies. GPCRs have also been notoriously difficult to crystallize owing to their intrinsic flexibility and the above mentioned reasons [6]. For such special cases, homology modeling aids in developing three dimensional models of such proteins. This has been possible through the understanding about the structure of GPCRs facilitated by homology modeling. Since many of these receptors lack experimentally solved structures, in silico methods like homology modeling were applied to gain insights. Template structure with high homology was used for modeling the structures to gain more advance insights on their function. Approximately one-fifth of the total GPCRs structure are solved whereas the remaining GPCR structures can be predicted by homology modeling. Three dimensional model building with the help of template helps us to predict protein structural and functional domains which further aids in drug discovery.

.This chapter deals with the contribution of homology modeling to the structural studies of GPCRs.

2. The importance and multifaceted functionality of GPCRs

The importance of G-protein coupled receptors (GPCRs) in the fields of biology, medicine and pharmaceutical studies have been extensively studied, well established and properly documented [7]. Due to its significance in playing a crucial role in various normal and pathological processes, GPCRs have become a major field of advanced research and a promising focus for drug discovery processes. The GPCRs have an extensive medical significance owing to their position and function within the human cell spanning the whole cell's plasma membrane. By this way it bridges the extra- and an intracellular environment which enables the GPCRs to act as signal transducers wherein it acclaims a direct mechanism for the transduction of extracellular messages into intracellular responses. In this way and together with their transmitters and effectors, GPCR systems function to modulate a broad spectrum of cellular phenomena dictated by the needs of the tissues and organs they serve. The gradient of GPCR distribution across vast majority of the body's organs and tissues and its primary role as signal transducers like converting transduce extracellular stimuli into intracellular signals at cellular levels makes it fascinating molecules from the perspective of advanced structural research.

Other fascinating roles of GPCRs include modulation of neuronal firing, regulation of ion transport across the plasma membrane and within intracellular organelles, modulation of homeostasis, control of cell division/proliferation, and modification of cell morphology. GPCRs are also an important target for cardiac drug therapy as decades of research revealed that GPCRs are the epicenters of many of the multiple causative factors of cardiovascular diseases like diabetes, obesity, environmental stressors and genetic factors [8]. Thus understanding the GPCR signaling mechanism in a healthy and an ailing heart may give better insights into treating cardiovascular problems.

There are over 200 cardio GPCRs and understanding their structural and functional properties is a key element in understanding the occurrence of heart diseases [9]. G-proteins consist of α, β, and γ subunits and a lot of global research has been carried out to check the various GPCR signaling pathways in a healthy and an ailing heart. Clinically targeted cardiac GPCRs like adrenergic receptors are responsible for translating chemical messages from the sympathetic nervous system into cardiovascular responses. Other such potentially targeted clinical GPCRs include angiotensin, endothelin, and adenosine receptors. Thus to study deeper about such cardio GPCRs one has to have structural studies carried out prior to analyzing its functionality.

Chemokine receptors belonging to the class A of GPCRs are involved in variety of physiologic functions, mostly related to the homeostasis of the immune system. They are also involved in multiple pathologic processes, including immune and autoimmune diseases, as well as cancer.

Other ailments caused when fundamental pathways governed by GPCRs go awry are asthma and strokes and cerebral hypoperfusion [10]. GPCRs control airway smooth muscle (ASM) contraction and increased airway resistance when coupled to Gq receptors. Airway epithelium and hematopoietic cells that are involved in control of lung inflammation that causes most asthma, have various pathways that are mediated by GPCRs. Arrestins regulate GPCR signaling and once again structural insights into the GPCRs is essential in understanding vital role of arrestins in those GPCR-mediated airway cell functions that are dysregulated in asthma.

3. A brief history on the structural study of GPCRs

GPCRs have been considered as one of the most desirable drug targets for the past few decades and have been investigated extensively. But the three dimensional structures of GPCRs have only recently become available. The first step in the structural study of GPCRs happened in the year 2000 with the initial crystal structure determination of Bovine rhodopsin (PDB: 1F88) through X-ray diffraction method [11]. The GPCR rhodopsin was purified from bovine rod outer segment (ROS) membranes. Multiwavelength anomalous diffraction (MAD) methods were employed to get the phasing information and the diffraction data from the crystallized Bovine rhodopsin were collected to 2.8 Å after mercury soaking. This experimental model of rhodopsin became a structural template for other GPCRs owing to the molecular size of Bovine rhodopsin, 348 amino acids, which was intermediate among the members of the GPCR family and thus can feature most of the essential parts of functional importance in G-protein activation.

An year later in 2001 the solution NMR method was used to solve the structure of Bovine rhodopsin (PDB ID: 1JFP [12]). It then took 7 long years to crystallize the next GPCR ADRB2 (PDB ID: 2RH1, 2R4R/2R4S [13, 14])[15]. It was solved using the LCP method that provides a more native, lipid environment for crystallization to a resolution of 2.4 Å. This delay was due to the need of numerous technological advancements required to crystallize membrane proteins like GPCRs. Developments in protein engineering, computational methods like homology modeling and heterologous protein expressions have accelerated structural determination of GPCRs [16]. In this structure solved via the LCP technique, the proteins are placed in a membrane-like environment where they can diffuse and interact with each other to form crystal lattice contacts on both complementary hydrophobic and hydrophilic regions. These structures served as the template for the other crystal structures that were solved afterwards. Other receptors like H1R, D3R and 5-HT1B belonging to the Rhodopsin family of GPCRs were solved in the following years and served as templates for all the other GPCR structures that were predicted by homology modeling in the following years of research.

Another important subfamily of class A GPCRs with a number of key physiologic roles are the Chemokine receptors [17]. So far (till 2020) only 5 different chemokine receptor complexes have had their crystal structure solved by researchers and they are CXCR4 [18] (PDB IDs: 3ODU, 3OE0, 3OE6, 3OE8, 3OE9, and 4RWS [18, 19]), CCR5 [20] (PDB IDs: 4MBS, 5UIW, 6AKX, and 6AKY [21–23]), US28 [24] (PDB IDs: 4XT1, 4XT3, 5WB1, and 5WB2 [24, 25]), CCR2 [26] (PDB IDs: 5T1A, 6GPS, and 6GPX [26, 27]) and CCR9 [28] (PDB ID: 5LWE [28]). Structure based drug design was the key in solving crystal structures of Chemokine receptors and its potential is reflected by the large amount of ligands found for various chemokine receptors. SBDD methods prove to be more effective when a crystal structure is available as homology models.

In 2011, Kobilka achieved another break-through when he and his team captured an image of the β-adrenergic receptor at the exact moment that it is activated by a hormone and sends a signal into the cell. This image is a molecular masterpiece [29]. This was the first step in the path that earned Brian Kobilka the Nobel Prize in Chemistry in the year 2012 for his groundbreaking discoveries about GPCRs along with Robert Lefkowitz. In the year 2011 and 2013, the first secretin family GPCR structure was solved (PDB ID: 4L6R, 4K5Y [30, 31]) and in the following year the first glutamate family GPCR structure was deposited in PDB (PDB ID: 4OR2, 4OO9 [32, 33]).

There are various databases available exclusively for GPCR structures like GPCR-EXP [https://zhanglab.ccmb. med.umich.edu/GPCR-EXP/] (database for experimentally solved GPCR structures) and GPCRdb [34] (web tools and diagrams that aid GPCR research) that profusely help the researchers. According to GPCR-EXP statistics there are 389 structures for 67 GPCRs belonging to different species deposited in the PDB. **Figure 2** gives us details about the total number of new experimental

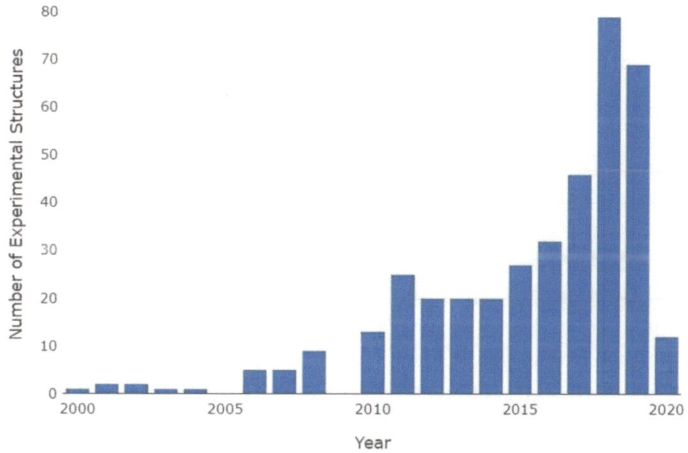

Figure 2.
The total number of new GPCR experimental structures available each year given by GPCR EXP database.

structures of GPCRs solved every year as recorded by GPCR-EXP database. There are still many more GPCR structures that are yet to be solved and these remain as an unturned page in the global research of GPCRs.

4. Role of homology modeling in unraveling the structures of GPCRs: a success story

Protein based virtual screening requires knowledge of three dimensional structure of targets. Researchers will have to face an overwhelming number of potential targets like GPCRs for which no or very few experimental 3-D information is available. Therefore, it is crucial in the near future to be able to use not only X-ray or NMR structures, but also GPCR models for protein-based virtual screening of chemical libraries. There are a lot of difficulties in obtaining significant amounts of pure and active recombinant GPCRs and this has been a huge problem in generating a lot of high resolution three dimensional structures of GPCRs. Low resolution GPCR structures of either bacteriorhodopsin or Bovine rhodopsin have paved way for many GPCR models. These models proved to be ineffective as they were not reliable enough for structure-based ligand design. The solving of crystal structure of the inactive dark-state rhodopsin back in the year 2001 was a huge mile stone in the structural study of GPCRs as a number of homology models of other class A GPCRs have been reported since then based on this structure.

Generally the importance of crystal structures is that they are useful to map sequence differences and to help analyze if the ortholog variant may affect the ligand binding and signaling of that particular GPCR. The first prerequisite for experimentally solving a protein structure is obtaining large amounts of stable, purified, homogeneous protein which can be used as templates to build a homology model. By means of in silico methods like homology modeling, crystal structures can be used to predict the effect of such ortholog variants. First step in developing homology models is the alignment of fingerprint motifs that are common among the family which are then are extrapolated to assign coordinates for the entire helical bundle. On the basis of databases of loop conformations and based on the specific application loop regions are either ignored or modeled accordingly [35]. As the template and query sequences used in homology modeling both belong to the GPCR family, the seven transmembrane (TM) helixes were properly transformed in the models according to that of the template structure. The RMSD between the model and the template structure must always preferably be less than range of 3 Å. Further the models were validated with the help of ERRAT plot [36], PROCHECK [37] and VERIFY3D [38].

One of the test case wherein homology modeling proved to be effective with the structural studies of GPCRs is the work done by Bissantz et al. where 3-D models of the D3, β2, and δ-opioid receptors were generated for future agonist screening as already several full agonists were known for each of these GPCRs [39]. Many GPCR models were set up to speculate if the "activated state" of GPCRs was conformationally more flexible than the antagonist-bound ground state. Apomorphine and pergolide (D3 receptor), epinephrine, and nylidrine (2 receptor), SNC-80, and TAN67 (−opioid receptor) were the agonists used for the refinement and two agonist-bound models were built for each receptor. An alternative activated-state model was also generated by substituting the single ligand-biased receptor to do comparative studies. When the amino acid sequences of the target receptors were aligned to the sequence of the Bovine rhodopsin template, the alignment coincided with the known structural features of GPCRs. It was observed that despite the low sequence identity when taking the whole TM sequence in account, the structurally and functionally important amino acids were highly conserved or compensated by amino acids of high similarity. These GPCR models are static though proteins are in reality more or less flexible which gave rise to more problems associated with docking. GPCR models based on a template with an identity of 20–30% can be expected to be of higher accuracy than when modeling other type of proteins based on a template with low-sequence identity and this was proven to be rue in this test case where in the antagonist-bound state models of three human GPCRs were proven to be suitable for virtual screening of GPCR antagonists. Although single template based models were seen to be less reliable. This was because all GPCR models that were used as templates have been derived from the inactive state of Bovine rhodospin, which was closer to an "antagonist-bound state" than to an "agonist-bound state" of the target GPCR and though their active site can be expanded the following conformational changes occurring in the receptor activation process could not be stimulated. A similar unreliability with the single template model was observed with all the GPCR homology models developed based on β2AR. Many models exist for β2AR, some of which have been improved upon with supporting biochemical data. All of these models were more similar to rhodopsin than β2AR. This was mainly because they were all homology models generated from single structural templates. The addition of multiple structural templates and conformational states to the pool of information on GPCRs later paved way to a new generation of more potent therapeutics targeting GPCR family. It is also not conclusive to come to a judgment where this unreliability of single template modeling stands strong as these modeling were conducted at a time when there were only few templates available. Judith Varady et al. used Bovine rhodopsin template to build the model of dopamine 3 (D3) subtype receptor which is a promising lead in treating drug addictions. The transmembrane helical region of the D3 receptor was modeled using Bovine rhodopsin template includes the ligand-binding site and showed sequence identity in the twilight region during homology modeling (sequence identity of 28%) [40].

Three-dimensional model of the human CCR5 receptor was developed by Fano A. et al. using a homology-based approach starting from the X-ray structure of the bovine rhodopsin receptor [41]. The reliability of these models was accessed using molecular docking and molecular dynamics studies. During this work there was no experimentally solved three dimensional chemokine receptor structures available and hence became a major hurdle in the deeper researches on the structural properties of these receptors. Therefore main ways to investigate the properties of CCR5 were homology modeling studies along with site-directed mutagenesis (SDM). Therefore a new model of CCR5 was built after consolidating all the information from the previously built models and also incorporating extensive molecular dynamics simulations (MD). Furthermore, flexible docking of a synthetic antagonist TAK779 and a novel docking protocol for natural agonists RANTES and MIP-1β was employed to develop the CCR5 models. The first crystal structure of bovine rhodopsin by Palczewski et al. served as the perfect template to build this model as the sequence identity increased to ~30% from previously being less than 20% when considering only the transmembrane helices (TMHs), and several of the amino acid residues essential for maintaining CCR5's architecture and receptor function were highly conserved. Pair wise alignment between the template and human CCR5 was carried out in CLUSTAL W and it was found that the anti-parallel β sheet loop of the second extracellular loop (ECL2) had higher sequence homology to the template. Out of the four cysteines which form two disulfide links in CCR5, Cys101-Cys178 had the anti-parallel β sheet loop of ECL2 and thus this loop was constructed by homology from the template structure using MODELER 6.2 [42]. The Cα Cartesian coordinates of the seven transmembrane helices and ECL2 were copied from the corresponding template (PDB: 1F88) and the N-terminal domain and the remaining loops were built de novo using MODELER 6.2. Confirmations of the models were done using PROCHECK and were selected as the input structure for MD Loop Refinement. The resulting model consisting of the TMHs and all ECLs and ICLs, was validated by MD conformational analysis, which showed it to be consistent with the then currently available SDM data and was used to gain insights into the molecular basis of the initiation and development of HIV-1 infection. This information could be useful in the rational design of HIV-1 entry blockers.

Chronologically, the time when the structural information about chemokine receptors was unavailable Gugan et al. in 2012 carried out the investigations on the binding site of CCR2 [43]. A comparative model was generated using the template structure of CXCR4 (PDB ID: 3ODU [44]). The structure of CXCR4 (PDB ID: 3ODU) was elucidated in 2010. One of the key findings along with the binding site residues is that the disulfide bridge was produced between Cys113-Cys190 of the selected CCR2 model and was also later observed in the crystal structure which was elucidated in 2016 (PDB ID: 5T1A [45]).

In the similar manner, Changdev et al. in 2013 developed the 3-D model for CCR5 using the template CXCR4 (PDB ID: 3ODU; resolution 2.5 Å) modeled by MODELER 9.2 [46] to explore the biding site of the receptor [47]. Significantly, the modeled structure coincides with the crystal structure of the CCR5 (PDB ID:4MBS [20]) whose structural information was determined by Tan et al. in 2013 [48].

The research by Anand et al. in 2011 on the accuracy of homology modeling revealed the comparison study between the reported models along with the crystal structure of CCR5 (PDB ID:4MBS) [49]. The findings have identified the importance of multi-template model in determining the insights of structural information of the receptor possessing its own merits and demerits. The inhibitor Maraviroc was docked to the single template and multi-template models of bovine rhodopsin (PDB ID: 1F88), β2 adrenergic receptor (PDB ID: 2RH1 [50]) and CXCR4 (PDB ID: 3ODU). The critical salt-bridge interaction established by Maraviroc with Glu283 of the receptor was genuinely observed in modeled structure and crystal structure.

In the process of building model of a particular GPCR usually many models are constructed with varying side chains and almost identical backbone. This is done to check which model among all the constructed models shows maximum affinity towards various ligands. So the model showing consisting binding mode is selected for further analysis. An example of this is the study done by Mateusz N et al. where 400 homology models of serotonin 5-HT1A receptor, one of the most documented monoamine GPCR, was modeled using Modeler 7v7 [51] with the crystal structure of bovine rhodopsin (PDB:1F88) as template [52]. These models varied considerably in their side chain but the polypeptide backbone varied only marginally from the template. Arylpiperazines test ligands were docked to all the 400 models with default parameters without any constraints. A detailed analysis of the docking poses revealed intrinsic information about crucial ionic bonds that were formed d almost exclusively in the case of receptors with the gauche(−) conformation of the Asp3.32 ø1 angle. Such insights led to the development of 200 new homology models with all the changes incorporated. Molecular docking was once again done on all the 200 new models and the complexes were scored using various scoring functions to choose the best models.

The past few years have seen remarkable advances in the structural biology of G-protein coupled receptors (GPCRs) and separate databases exist to study GPCRs. The applications of structural studies of GPCRs have various goals and these goals trigger myriad scientific investigations. For the GPCRs whose structures have now been solved, the homology models developed earlier based on rhodopsin, have been the first step in discovering the versatility of their structural studies. Due to the increase in the available GPCR structures, the templates used to build the structure for homology modeled GPCRs show a drastic increase in similarity and query coverage in the recent years. This enhances the structure of the models which are being constructed with the upcoming elucidated structures of GPCRs.

5. Conclusion

The research in GPCRs is a global phenomenon and this is possible only if we have structural insights based on structural studies of GPCRs. Owing to the difficulty in crystallizing the GPCRs, it was once construed that structural studies of GPCRs were impossible. But with the technological advancements in the computational techniques, building a model structure based on the homology of a particular receptor with a template structure became possible. Thus homology modeling and models generated via tools like MODELER unraveled the unexplored arenas in the research of GPCRs. These models served a greater purpose to the pharmaceutical industries wherein GPCRs became famous drug discovery targets. The many experimental structures constructed using previously solved structures as templates were further scrutinized based on their efficiency in showing a consisting binding mode with various ligands. Recent times have seen use of Cryo EM techniques in solving structures of GPCRs. But still contribution made by techniques like homology modeling in the structural studies of GPCRs will always remain as a mile stone.

Conflict of interest

The authors declare no conflict of interest.

Author details

Ananthasri Sailapathi[1], Seshan Gunalan[1], Kanagasabai Somarathinam[1], Gugan Kothandan[1]* and Diwakar Kumar[2]

1 Biopolymer Modeling Laboratory, CAS in Crystallography and Biophysics, University of Madras, Chennai, India

2 Department of Microbiology, Assam University, Assam, India

*Address all correspondence to: drgugank@gmail.com

References

[1] Vyas VK, Ukawala RD, Ghate M, Chintha C. Homology modeling a fast tool for drug discovery: current perspectives. Indian journal of pharmaceutical sciences. 2012 Jan;74(1):1.DOI: 10.4103/0250-474X.102537

[2] White SH. Biophysical dissection of membrane proteins. Nature. 2009 May 20;459(7245):344.DOI: 10.1038/nature0814

[3] Carpenter EP, Beis K, Cameron AD, Iwata S. Overcoming the challenges of membrane protein crystallography. Current opinion in structural biology. 2008 Oct 1;18(5):581-6.DOI: 10.1016/j.sbi.2008.07.001

[4] Kroeze WK, Sheffler DJ, Roth BL. G-protein-coupled receptors at a glance. Journal of cell science. 2003 Dec 15;116(24):4867-9. DOI:10.1242/jcs.00902

[5] Rosenbaum DM, Rasmussen SG, Kobilka BK. The structure and function of G-protein-coupled receptors. Nature. 2009 May;459(7245):356-63.DOI: 10.1038/nature08144.

[6] Zhang D, Zhao Q, Wu B. Structural studies of G protein-coupled receptors. Molecules and cells. 2015 Oct 31;38(10):836.DOI: 10.14348/molcells.2015.0263

[7] Salon JA, Lodowski DT, Palczewski K. The significance of G protein-coupled receptor crystallography for drug discovery. Pharmacological reviews. 2011 Dec 1;63(4):901-37.DOI: 10.1124/pr.110.003350

[8] Fernandez-Patron C, Filep JG. GPCRs in cardiovascular pathologies. Drug discovery today. Disease mechanisms. 2012;9(3):e75.DOI: 10.1016/j.ddmod.2012.07.001

[9] Salazar NC, Chen J, Rockman HA. Cardiac GPCRs: GPCR signaling in healthy and failing hearts. Biochimica et Biophysica Acta (BBA)-Biomembranes. 2007 Apr 1;1768(4):1006-18. DOI:10.1016/j.bbamem.2007.02.010

[10] Penn RB, Bond RA, Walker JK. GPCRs and arrestins in airways: implications for asthma. InArrestins-Pharmacology and Therapeutic Potential 2014 (pp. 387-403). Springer, Berlin, Heidelberg.DOI: 10.1007/978-3-642-41199-1_20

[11] Palczewski K, Kumasaka T, Hori T, Behnke CA, Motoshima H, Fox BA, Le Trong I, Teller DC, Okada T, Stenkamp RE, Yamamoto M. Crystal structure of rhodopsin: AG protein-coupled receptor. science. 2000 Aug 4;289(5480):739-45. DOI: 10.1126/science.289.5480.739

[12] Yeagle PL, Choi G, Albert AD. Studies on the structure of the G-protein-coupled receptor rhodopsin including the putative G-protein binding site in unactivated and activated forms. Biochemistry. 2001 Oct 2;40(39):11932-7.DOI: 10.1021/bi015543f

[13] Cherezov V, Rosenbaum DM, Hanson MA, Rasmussen SG, Thian FS, Kobilka TS, Choi HJ, Kuhn P, Weis WI, Kobilka BK, Stevens RC. High-resolution crystal structure of an engineered human β2-adrenergic G protein–coupled receptor. science. 2007 Nov 23;318(5854):1258-65.DOI: 10.2210/pdb2RH1/pdb

[14] Rasmussen SG, Choi HJ, Rosenbaum DM, Kobilka TS, Thian FS, Edwards PC, Burghammer M, Ratnala VR, Sanishvili R, Fischetti RF, Schertler GF. Crystal structure of the human β 2 adrenergic G-protein-coupled receptor. Nature. 2007

Nov;450(7168):383-7.DOI: 10.1038/nature06325

[15] Yeagle PL, Choi G, Albert AD. Studies on the structure of the G-protein-coupled receptor rhodopsin including the putative G-protein binding site in unactivated and activated forms. Biochemistry. 2001 Oct 2;40(39):11932-7. DOI: 10.1021/bi015543f

[16] Gacasan SB, Baker DL, Parrill AL. G protein-coupled receptors: the evolution of structural insight. AIMS biophysics. 2017;4(3):491.DOI: 10.3934/biophy.2017.3.491

[17] Arimont M, Hoffmann C, de Graaf C, Leurs R. Chemokine Receptor Crystal Structures: What Can Be Learned from Them?. Molecular pharmacology. 2019 Dec 1;96(6):765-77. DOI: 10.1124/mol.119.117168

[18] Wu B, Chien EY, Mol CD, Fenalti G, Liu W, Katritch V, Abagyan R, Brooun A, Wells P, Bi FC, Hamel DJ. Structures of the CXCR4 chemokine GPCR with small-molecule and cyclic peptide antagonists. Science. 2010 Nov 19;330(6007):1066-71.DOI: 10.1126/science.1194396

[19] Qin L, Kufareva I, Holden LG, Wang C, Zheng Y, Zhao C, Fenalti G, Wu H, Han GW, Cherezov V, Abagyan R. Crystal structure of the chemokine receptor CXCR4 in complex with a viral chemokine. Science. 2015 Mar 6;347(6226):1117-22.DOI: 10.1126/science.1261064

[20] Tan Q, Zhu Y, Li J, Chen Z, Han GW, Kufareva I, Li T, Ma L, Fenalti G, Li J, Zhang W. Structure of the CCR5 chemokine receptor–HIV entry inhibitor maraviroc complex. Science. 2013 Sep 20;341(6152):1387-90.DOI: 10.1126/science.1241475

[21] Tan Q, Zhu Y, Li J, Chen Z, Han GW, Kufareva I, Li T, Ma L, Fenalti G, Li J, Zhang W. Structure of the CCR5 chemokine receptor–HIV entry inhibitor maraviroc complex. Science. 2013 Sep 20;341(6152):1387-90.DOI: 10.1126/science.1241475

[22] Zheng Y, Han GW, Abagyan R, Wu B, Stevens RC, Cherezov V, Kufareva I, Handel TM. Structure of CC chemokine receptor 5 with a potent chemokine antagonist reveals mechanisms of chemokine recognition and molecular mimicry by HIV. Immunity. 2017 Jun 20;46(6):1005-17. DOI: 10.1016/j.immuni.2017.05.002

[23] Peng P, Chen H, Zhu Y, Wang Z, Li J, Luo RH, Wang J, Chen L, Yang LM, Jiang H, Xie X. Structure-based design of 1-heteroaryl-1, 3-propanediamine derivatives as a novel series of CC chemokine receptor 5 antagonists. Journal of medicinal chemistry. 2018 Sep 20;61(21):9621-36.DOI: 10.1021/acs.jmedchem.8b01077

[24] Burg JS, Ingram JR, Venkatakrishnan AJ, Jude KM, Dukkipati A, Feinberg EN, Angelini A, Waghray D, Dror RO, Ploegh HL, Garcia KC. Structural basis for chemokine recognition and activation of a viral G protein–coupled receptor. Science. 2015 Mar 6;347(6226):1113-7.DOI: 10.1126/science.aaa5026

[25] Miles TF, Spiess K, Jude KM, Tsutsumi N, Burg JS, Ingram JR, Waghray D, Hjorto GM, Larsen O, Ploegh HL, Rosenkilde MM. Viral GPCR US28 can signal in response to chemokine agonists of nearly unlimited structural degeneracy. Elife. 2018 Jun 8;7:e35850.DOI: 10.7554/eLife.35850

[26] Zheng Y, Qin L, Zacarías NV, de Vries H, Han GW, Gustavsson M, Dabros M, Zhao C, Cherney RJ, Carter P, Stamos D. Structure of CC chemokine receptor 2 with orthosteric and allosteric antagonists. Nature. 2016 Dec;540(7633):458-61.DOI: 10.1038/nature20605

[27] Apel AK, Cheng RK, Tautermann CS, Brauchle M, Huang CY, Pautsch A, Hennig M, Nar H, Schnapp G. Crystal structure of CC chemokine receptor 2A in complex with an orthosteric antagonist provides insights for the design of selective antagonists. Structure. 2019 Mar 5;27(3):427-38.DOI: 10.1016/j.str.2018.10.027

[28] Oswald C, Rappas M, Kean J, Doré AS, Errey JC, Bennett K, Deflorian F, Christopher JA, Jazayeri A, Mason JS, Congreve M. Intracellular allosteric antagonism of the CCR9 receptor. Nature. 2016 Dec;540(7633):462-5.DOI: 10.1038/nature20606

[29] Cherezov V, Rosenbaum DM, Hanson MA, Rasmussen SG, Thian FS, Kobilka TS, Choi HJ, Kuhn P, Weis WI, Kobilka BK, Stevens RC. High-resolution crystal structure of an engineered human β2-adrenergic G protein–coupled receptor. science. 2007 Nov 23;318(5854):1258-65.DOI: 10.1126/science.1150577

[30] Siu FY, He M, De Graaf C, Han GW, Yang D, Zhang Z, Zhou C, Xu Q, Wacker D, Joseph JS, Liu W. Structure of the human glucagon class B G-protein-coupled receptor. Nature. 2013 Jul;499(7459):444-9.DOI: 10.1038/nature12393

[31] Hollenstein K, Kean J, Bortolato A, Cheng RK, Doré AS, Jazayeri A, Cooke RM, Weir M, Marshall FH. Structure of class B GPCR corticotropin-releasing factor receptor 1. Nature. 2013 Jul;499(7459):438-43. DOI: 10.1038/nature12357

[32] Wu H, Wang C, Gregory KJ, Han GW, Cho HP, Xia Y, Niswender CM, Katritch V, Meiler J, Cherezov V, Conn PJ. Structure of a class C GPCR metabotropic glutamate receptor 1 bound to an allosteric modulator. Science. 2014 Apr 4;344(6179):58-64. DOI: 10.1126/science.1249489

[33] Doré AS, Okrasa K, Patel JC, Serrano-Vega M, Bennett K, Cooke RM, Errey JC, Jazayeri A, Khan S, Tehan B, Weir M. Structure of class C GPCR metabotropic glutamate receptor 5 transmembrane domain. Nature. 2014 Jul;511(7511):557-62.DOI: 10.1038/nature13396

[34] Munk C, Isberg V, Mordalski S, Harpsøe K, Rataj K, Hauser AS, Kolb P, Bojarski AJ, Vriend G, Gloriam DE. GPCRdb: the G protein-coupled receptor database–an introduction. British Journal of Pharmacology. 2016 Jul;173(14):2195-207.DOI: 10.1111/bph.13509

[35] Castleman PN, Sears CK, Cole JA, Baker DL, Parrill AL. GPCR homology model template selection benchmarking: Global versus local similarity measures. Journal of Molecular Graphics and Modelling. 2019 Jan 1;86:235-46. DOI: 10.1016/j.jmgm.2018.10.016

[36] Colovos C, Yeates TO. ERRAT: an empirical atom-based method for validating protein structures. Protein Sci. 1993;2(9):1511-9.DOI: 10.1002/pro.5560020916

[37] Laskowski RA, MacArthur MW, Moss DS, Thornton JM. PROCHECK: a program to check the stereochemical quality of protein structures. Journal of applied crystallography. 1993 Apr 1;26(2):283-91.DOI: 10.1107/S0021889892009944

[38] Eisenberg D, Lüthy R, Bowie JU. VERIFY3D: assessment of protein models with three-dimensional profiles. InMethods in enzymology 1997 Jan 1 (Vol. 277, pp. 396-404). Academic Press. DOI: 10.1016/s0076-6879(97)77022-8

[39] Bissantz C, Bernard P, Hibert M, Rognan D. Protein-based virtual

screening of chemical databases. II. Are homology models of G-protein coupled receptors suitable targets?. Proteins: Structure, Function, and Bioinformatics. 2003 Jan 1;50(1):5-25. DOI: 10.1002/prot.10237

[40] Varady J, Wu X, Fang X, Min J, Hu Z, Levant B, Wang S. Molecular modeling of the three-dimensional structure of dopamine 3 (D3) subtype receptor: discovery of novel and potent D3 ligands through a hybrid pharmacophore-and structure-based database searching approach. Journal of medicinal chemistry. 2003 Oct 9;46(21):4377-92.DOI: 10.1021/jm030085p

[41] Fano A, Ritchie DW, Carrieri A. Modeling the structural basis of human CCR5 chemokine receptor function: from homology model building and molecular dynamics validation to agonist and antagonist docking. Journal of chemical information and modeling. 2006 May 22;46(3):1223-35.DOI: 10.1021/ci050490k

[42] Sánchez R, Šali A. Comparative protein structure modeling: introduction and practical examples with modeller. InProtein Structure Prediction 2000 (pp. 97-129). Humana Press. DOI: 10.1385/1-59259-368-2:97

[43] Kothandan G, Gadhe CG, Cho SJ. Structural insights from binding poses of CCR2 and CCR5 with clinically important antagonists: a combined in silico study. PloS one. 2012 Mar 27;7(3):e32864. DOI: 10.1371/journal.pone.0032864

[44] Wu B, Chien EY, Mol CD, Fenalti G, Liu W, Katritch V, Abagyan R, Brooun A, Wells P, Bi FC, Hamel DJ. Structures of the CXCR4 chemokine GPCR with small-molecule and cyclic peptide antagonists. Science. 2010 Nov 19;330(6007):1066-71.DOI: 10.1126/science.1194396

[45] Zheng Y, Qin L, Zacarías NV, de Vries H, Han GW, Gustavsson M, Dabros M, Zhao C, Cherney RJ, Carter P, Stamos D. Structure of CC chemokine receptor 2 with orthosteric and allosteric antagonists. Nature. 2016 Dec;540(7633):458-61.DOI: 10.1038/nature20605

[46] Šali A, Blundell TL. Comparative protein modelling by satisfaction of spatial restraints. Journal of molecular biology. 1993 Dec 5;234(3):779-815. DOI: 10.1006/jmbi.1993.1626

[47] Gadhe CG, Kothandan G, Cho SJ. Computational modeling of human coreceptor CCR5 antagonist as a HIV-1 entry inhibitor: using an integrated homology modeling, docking, and membrane molecular dynamics simulation analysis approach. Journal of Biomolecular Structure and Dynamics. 2013 Nov 1;31(11):1251-76. DOI: 10.1080/07391102.2012.732342

[48] Tan Q, Zhu Y, Li J, Chen Z, Han GW, Kufareva I, Li T, Ma L, Fenalti G, Li J, Zhang W. Structure of the CCR5 chemokine receptor–HIV entry inhibitor maraviroc complex. Science. 2013 Sep 20;341(6152):1387-90.DOI: 10.1126/science.1241475

[49] Aanand V, Anitha R, Kanagasabai S, Seshan G, Gadhe CG, Kothandan G. Unveiling the Accuracy of Homology Modeling to Elucidate the Structure of GPCRs-HIV Co-receptor-CCR5 as a Case Study. Letters in Drug Design & Discovery. 2018 Oct 1;15(10):1068-78. DOI: 10.2174/15701808146661711 20153956

[50] Cherezov V, Rosenbaum DM, Hanson MA, Rasmussen SG, Thian FS, Kobilka TS, Choi HJ, Kuhn P, Weis WI, Kobilka BK, Stevens RC. High-resolution crystal structure of an engineered human β2-adrenergic G protein–coupled receptor. science. 2007 Nov 23;318(5854):1258-65.DOI: 10.1126/science.1150577

[51] Šali A, Blundell TL. Comparative protein modelling by satisfaction of spatial restraints. Journal of molecular biology. 1993 Dec 5;234(3):779-815. DOI: 10.1006/jmbi.1993.1626

[52] Nowak M, Kołaczkowski M, Pawłowski M, Bojarski AJ. Homology modeling of the serotonin 5-HT1A receptor using automated docking of bioactive compounds with defined geometry. Journal of medicinal chemistry. 2006 Jan 12;49(1):205-14. DOI: 10.1021/jm050826h

Chapter 5

Homology Modeling of Tubulin Isotypes to Investigate MT-Tau Interactions

Vishwambhar Vishnu Bhandare

Abstract

The Homology modeling techniques uses the template structure(s) to model the full-length structure of unknown sequence. It is being used for determining the structure of biological macromolecules, especially proteins. The wide applications of homology modeling approach have helped us to address various challenging problems in the field of biological sciences and drug discovery despite the limitations in using analytical techniques like X-ray, NMR and CryoEM techniques. Here, this chapter emphasize on application of homology modeling in determining MT-Tau interactions which are important in the Alzheimer disease. In Alzheimer diseases, tau detaches from MTs in misfolded shape and forms insoluble aggregates in neurons due to post-translational modifications. MT-tau interactions are largely unknown due to differential expression of neuronal specific tubulin isotypes and intrinsically disordered nature of tau. MTs play crucial roles in important cellular functions including cell division, transport of vesicles, cell signaling, cell motility etc. MTs are composed of different tubulin isotypes which differs mainly at C-terminal tail. In humans, nine β-tubulin isotypes have been reported which are expressed differently in different tissues. Structures for different tubulin isotypes are still lacking due to their complex differential expression pattern and purification. Hence, homology modeling approach allowed us to generate homology models for different neuronal specific tubulin isotypes and study their interactions with tau repeats. It is believed that this study would gain more structural and functional insights to the linked Alzheimer diseases.

Keywords: homology modeling, microtubule, tubulin isotypes, Alzheimer disease, molecular modeling

1. Introduction

Bioinformatics is an interdisciplinary science that uses both computational and informational approaches to retrieve, analyze, organize, visualize, store and develop biological data [1]. It is widely applied in the field of life sciences, especially in functional genomics, sequence analysis, proteomics, drug discovery, etc. Prediction of the structure and functions of the genes and proteins have become a fundamental task in the life science researches. The present book chapter involves molecular modeling study to investigate intermolecular interactions between Microtubule (MT) and Tau. Though these interactions are important in Alzheimer's disease, the detailed knowledge on MT-Tau interactions are still lacking mainly because of two reasons (i) Lack of full-length structure of tau due to its intrinsically disordered nature and (ii) Differential expression of tubulin isotypes in different type of cells in particular brain and neuronal cells. Earlier experimental efforts have been made to elucidate these interactions using solution structure (PDB ID: 2MZ7.pdb) however correct binding mode and atomistic interactions at structural level are poorly understood. Therefore, this chapter focuses on application of molecular modeling techniques in understanding important MT-Tau interactions in the Alzheimers disease. Bioinformatics approaches like sequence analysis, homology modeling. MD simulations and binding energy calculations are employed systematically to address this challenging problem in the field of Alzheimer's disease.

Tau is intrinsically disordered protein encoded by '*mapt*' gene located on chromosome 17 [2]. The primary function of the tau protein is to bind and stabilize the microtubule. It is abundantly expressed in the brain and neuronal tissues hence its misregulation is associated with the Alzheimers and other neurodegenerative disorders [3, 4]. Till date about six isoforms of tau are reported in the human central nervous system. The length of these six isoforms varies between 352 to 441 residues [5].

Primary structure of tau contains the projectile domain at N-terminal (residue 1–244) which is composed of the acidic and proline-rich region, and the C-terminal repeat domain which consists of 4 repeats i.e., R1, R2, R3 and R4 (residues 245–441) (**Figure 1**). The six isoforms of tau mainly differs by the existence of either R3 or R4 repeats at the C-terminal domain [6]. The one of the tau isoforms is referred as longest isoform mostly observed in humans which comprises 4 repeats i.e., R1, R2, R3 and R4. while the shortest isoform of tau has only 3 repeats (R1, R2 and R3). This shortest isoform of tau is reported in the fetus brain and less common in adults [5, 7]. **Figure 1A** represents the structure of tau repeat region R2 which is bound to the MT composed of β/α/β tubulin subunits [8]. Hereafter, tau repeat R2 will be mentioned as 'TauR2' for the simplicity. The tau repeats R1, R2, R3 and R4 prefers to bind at the outer surface of microtubule (MT) to stabilize it (**Figure 1A**) and regulates MT polymerization [6]. **Figure 1B** represents domain organization in the tau primary structure and **Figure 1C** shows the sequence of TauR2 which is reported in the CryoEM model. It is well established that tau primarily helps in the assembly and stabilization of axonal MTs, which contributes to the proper functioning of neuronal cells [9]. However, recent studies have reported new functional role of the tau in addition to the axonal, i.e. labile domain of the MT to promote its assembly [10]. Tau detaches from the MTs and forms abnormal, fibrillar structures of insoluble aggregates due to post-translational modifications in Alzheimer diseases and other neurodegenerative diseases associated with tau [11, 12].

Full-length structure of tau protein is not yet determined using X-ray crystallographic techniques due to its intrinsically disordered nature. Also, the efforts to find its solution structure using NMR spectroscopy have failed [13]. Thus, the MT-Tau interactions have been studied so far using various biochemical and biophysical techniques [14–17]. The CryoEM have showed marginal success in determining the structure of tau repeat R2 bound to MT however it shows discontinues density of tau repeats along with each protofilament upon MT binding [8]. Hence, they synthetically developed R1 and R2 repeat of tau and their interactions with MT were examined. These two tau repeats adopts the extended conformation along the crest of protofilament which stabilizes the MT structure by binding to the interface of tubulin dimers [8].

MTs are made from αβ-tubulin heterodimer subunits [18]. In human, seven α-tubulin and nine β-tubulin isotypes are reported showing their tissue-specific expressions. For instance, βI tubulin isotype is ubiquitously expressed in all cells,

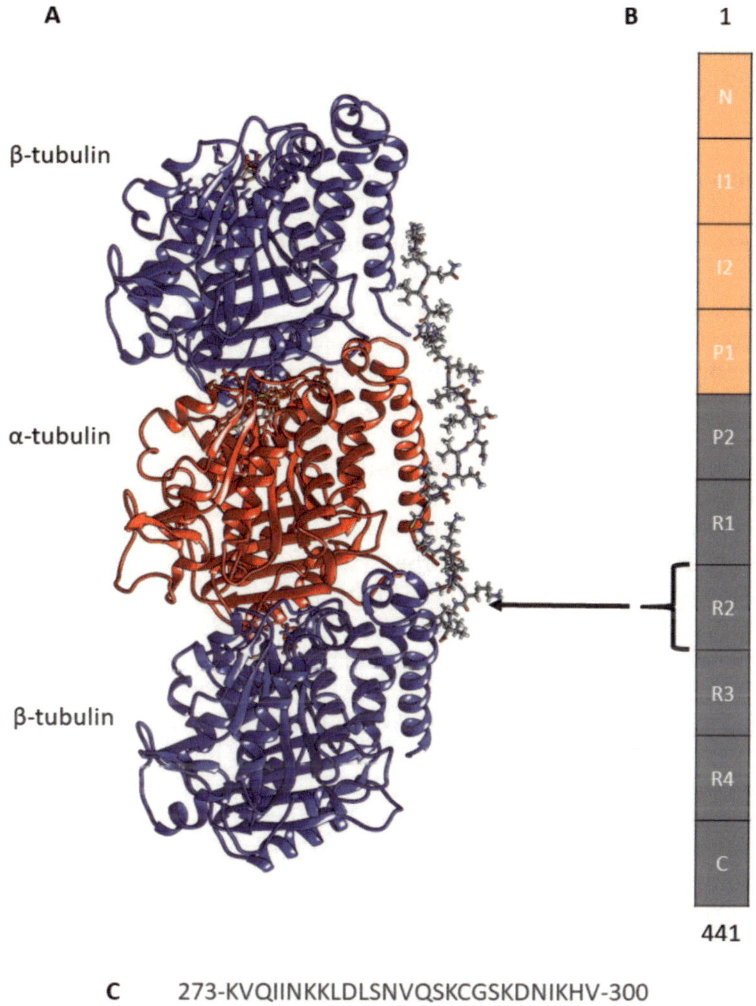

Figure 1.
CryoEM Structure of tubulin subunits bound to TauR2. (A) tubulin subunits bound to TauR2 in CryoEM structure 6CVN.pdb. TauR2 domain binds at the outer surface of the MT. (B) Domain organization in tau, (C) sequence of TauR2. [Source: Bhandare et al, 2019; doi: 10.1038/s41598-019-47249-7].

βII and βIII tubulin isotype are mainly expressed in brain and neuronal cells, and βVI tubulin isotype is expressed in erythroid cells and platelets [19]. The βI tubulin isotype reported to play crucial role in cell viability, βII tubulin isotype is important for neurite growth and βIII tubulin isotype protects nerve cell against free radicals and reactive oxygen species [20]. It has been well known that all β-tubulin isotypes share a significant residue conservation except the C-terminal tail region of MT [21–24] which is flexible in nature and structurally disordered. The C- tail region of all these isotypes overhang outwards of the MTs. The C-tail shows interactions with various MAPs including tau and regulate MT dynamics [25, 26].

It is well documented that the composition of β-tubulin isotypes (i) affects MT dynamic instability [27, 28], (ii) their interaction with motor proteins [29], (iii) their binding to the anti-drugs [21, 22, 30] and (iv) different MAPs including tau [31, 32]. These tubulin isotypes show tissue specific expression as their relative proportion varies greatly in different type of cells [20, 33, 34]. It is also well established that binding of tau to the MT promote or demote microtubule polymerization [35]. However, the differential binding affinity of tau to the various β-tubulin isotypes expressed in different types of cells is completely unknown. Therefore, we studied relative binding affinity of Tau repeat region R2 with neuronal specific β-tubulin isotypes namely βI, βIIb, and βIII using molecular modeling [36].

2. Methodology

2.1 Sequence analysis and homology modeling of tubulin isotypes

In biological research, sequence analysis plays a major role as it has wide range of applications such as, (i) whole genome sequencing and annotation, (ii) identification of functional elements in the sequence, (iii) gene prediction, (iv) comparative genomics, (v) protein classification, (vi) protein and RNA structure prediction, (vii) evolutionary studies, etc. Protein sequences reveal the evolutionary history and hence, the events occurred during evolutions can be traced from the protein sequences.

The structure 6CVN.pdb is used as template structure for homology modeling of neuronal specific human tubulin isotypes namely βI, βIIb and βIII tubulin (uniprot IDs Q9H4B7, Q9BVA1 and Q136509). The multiple sequence alignment of these sequence was performed using 'clustal omega' tool [37]. The multiple sequence alignment reveals that residue variations is mainly at the C-terminal tail when compared to the other regions of the protein. The high-resolution cryo-EM structure of β/α/β-tubulin bound with TauR2 was recently deposited in the RCSB structural databases [8] was used as a template to build βI/α/βI-TauR2, βIIb/α/βIIb-TauR2 and βIII/α/βIII-TauR2 complexes. The structure for C-terminal tail was absent in the template structure (6CVN.pdb), therefore the tail region was modeled using the Modeler 9v20. Hereafter, template structure with modeled C-terminal tail region would be referred as 6CVN* in the further discussion.

Template based homology models for neuronal specific tubulin isotypes βI, βIIb, βIII was build using Modeler 9v20 [38]. The least discrete optimized potential energy (DOPE) score model was selected for further use. The stereo-chemical properties of these modeled subunits were evaluated and validated using the GMQE score [39], verify3D [40], ERRAT score [41] and Ramachandran plot through PROCHECK [42]. The selected subunit models were further used to 6CVN*-TauR2, build βI/α/βI-TauR2, βIIb/α/βIIb-TauR2 and βIII/α/βIII-TauR2. These complexes namely 6CVN-TauR2, 6CVN*-TauR2, βI/α/βI-TauR2, βIIb/α/βIIb-TauR2 and βIII/α/βIII-TauR2 were further subjected for energy minimization to get their least energy state. Here we used Steepest Descent and Conjugate Gradient methods in Gromacs 2018.1 software for minimization [43]. The process of energy minimization is a numerical procedure aimed to find a minimum on the potential energy surface (PES) of the newly modeled conformation which mostly exists at a higher energy level. These minimized models were used as a starting input structures for molecular dynamics simulations to understand the binding mode and binding affinity of TauR2 towards neuronal specific tubulin isotypes βI, βII and βIII.

2.2 Molecular dynamics simulations of tubulin-TauR2 complexes

Molecular dynamics (MD) simulation plays a key role in exploring the structure and function of biological macromolecules [44]. In MD simulations, the dynamic behavior of the molecule is studied as a function of time. Molecular dynamics is being routinely used to address various biological problems such as biomolecular interactions (Protein–protein, protein-DNA/RNA), molecular pathways, Drug-receptor interactions, dynamics of protein folding, protein aggregations, protein structure prediction, etc. Tremendous development in high performance computing and simplicity of the basic MD algorithm has shortened the time required to perform molecular dynamics simulation and hence, studying larger systems became an easier task [45].

All atom molecular dynamics (MD) simulation was performed in explicit solvent on the modeled tubulin-TauR2 complexes (i.e. 6CVN-TauR2, 6CVN*-tau, βI/α/βI-TauR2, βIIb/α/βIIb-TauR2 and βIII/α/βIII-TauR2) using GROMACS 2018.1 [43, 46]. The 'Amber99SB-ILDN' force field [47] was chosen for the MD simulation because it is well customized to handle the parameters for GTP,GDP and MG atoms which are the functional players of all the modeled tubulin-TauR2 complexes. The force field parameters for the GDP and GTP molecules were retrieved from the amber parameter database [48, 49]. The '*xleap*' module of AmberTools was used to generate topology files and initial starting coordinates for all the complexes [50]. All the modeled tau-tubulin complexes were placed at the centre of a cubic shaped solvation box having dimension of 15 Å from the extent of the molecule and TIP3P water model was used for solvation. All the systems were neutralized by adding appropriate number of required counterions. The topology files generated using xleap module of AmberTools were converted to Gromacs compatible topologies with ParmEd tool [51]. Energy minimization was carried out in two steps, In the first step, steepest descent algorithm was used for 50,000 followed by the conjugate gradient [46]. The energy minimized models were equilibrated using canonical ensemble (NVT) followed by isothermal-isobaric ensemble (NPT) for 1 ns. In the NVT equilibration systems were heated to 300 K using V-rescale, a modified Berendsen thermostat [46]. These heated systems were further equilibrated using the Parrinello-Rahman barostat to maintain constant pressure of 1 bar. The production MD simulations were performed for 100 ns without restraining any atoms over all the tubulin-TauR2 complexes using parameters discussed in earlier study [52]. The PME method was used to treat long range electrostatic interactions [53, 54] and covalent bonds involving H-atoms were constrained by using 'LINCS' algorithm [55]. The 2 fs time step was set for integrating the newtonian equation during the MD simulation. Similar protocol was adopted to perform MD simulation on three additional systems (i) 6CVN* (without tau), (ii) free tau and (iii) 6CVN*-polyA (as negative control) having 27 amino acids residues. All the MD simulation trajectories were further analyzed by using the GROMACS 2018.1 inbuilt tools [43, 46]. The general parameters explaining the conformational stability such as Root Mean Square Deviation (RMSD), Root Mean Square Fluctuations (RMSF) and Radius of Gyration (Rg) were measured, and the equations used for calculation of these parameters are tabulated in the **Table 1**.

The primary sequence of a protein is a linear chain of amino acids linked by peptide bonds. There is a direct link between the protein sequence, structure and function. The secondary structure of a protein is comprised of coils, α-helices, β-sheets,

S. no.	Parameter	Equation	Component
1	RMSD	$RMSD = \sqrt{\frac{\sum_{i=1}^{N} d_i^2}{N}}$	*di* is the distance, 'N' is number of atoms
2	RMSF	$RMSF = \sqrt{\frac{\sum_{j=1}^{t} \left(x_i(t_j) - \bar{x}_i\right)^2}{t}}$	x_i is atom position at time t, reference position is $\bar{x}_i$
3	R_g	$R_g = \sqrt{\frac{\sum_{i=1}^{N} (x_i - x_{com})^2}{N}}$	x_{com} centre of mass, x_i is the distance at time t from their centre of mass

Table 1.
Equations used to calculate RMSD, RMSF and R_g.

β-bridge, bend, turn, coils, π-helix and 3_{10}-helices. DSSP is an algorithm developed by Kabsch and Sander to extract the secondary structural features based on atomic coordinates [56]. The overall stability of the structure is highly determined by the stable dynamics of these secondary structures and any significant changes in secondary structure attributes to the structural flexibility/fold as well as functional diversity of the protein. Hence, conformational changes in the secondary structure during MD simulation were analyzed using the DSSP programme [56]. The simulation movies over the entire trajectories were generated using the VMD software [57] and publication quality images were generated using the Biovia Discovery studio visualizer [58] and Chimera software [59].

2.3 Calculations of contact surface area (CSA) for tubulin-TauR2 complexes

Solvent Accessible Surface Area (SASA) is used to represent the degree of hydration of a biomolecule. SASA also be especially useful to quantify the stability of the biomolecular complexes in the aqueous medium. The C-terminal tail of the tubulin subunits is highly dynamic in nature and has no definite secondary structure, hence it affects the overall hydrophobic SASA. Therefore, interface of the MT (in this case tubulin trimer made up of β/α/β subunits) where TauR2 binds at the exterior surface has been selected for the calculating the precise CSA. The in-built gromacs tool "*gmx sasa*" [60] was used to calculate the SASA. In addition, SASA is also calculated for the tubulin subunits and the TauR2.

2.4 Binding affinity of tauR2 towards different neuronal specific tubulin isotypes

The biomolecular recognition pattern mainly depends on the binding ability of the interacting biomolecules. The binding affinity as well as the energy between the two interacting molecules can be calculated using various theoretical approaches like (i) Pathway methods such as Thermodynamic integration (TI) as well as Free energy perturbation (FEP) and (ii) End point methods such as Molecular Mechanics Poisson-Boltzman Surface Area (MM/PBSA) and Molecular Mechanics Generalized Born Surface Area (MM/GBSA) [61]. In the present study, MM/PBSA approach was used to calculate relative binding energies of the simulated molecules. This MMPBSA approach is very popular, computationally less expensive, and has better accuracy even for the larger systems [62].

Here, the binding affinity between different neuronal specific tubulin isotypes and TauR2 was estimated by performing relative binding energy calculation similar to earlier studies [63–65]. The stable trajectory observed in between 70 ns to 100 ns was chosen to perform the binding energy calculations for all the tubulin-TauR2 complexes. The 'g_mmpbsa' tool v1.6 was used to perform binding energy calculation using MM/PBSA approach [66]. The parameters for the binding energy calculations were chosen from the earlier similar studies [52, 65, 67–69]. In the MMPBSA methods binding energy (ΔG_{bind}) of tubulin and TauR2 was calculated by using the following Eq. (1),

$$\Delta G_{bind} = \Delta G_{tubulin-TauR2} - \left(\Delta G_{tubulin} + \Delta G_{TauR2}\right) \quad (1)$$

Where, the $\Delta G_{tubulin-TauR2}$, $\Delta G_{tubulin}$ and ΔG_{TauR2} represents the average free energies of the complex (tubulin-TauR2), receptor (tubulin) and ligand (TauR2), respectively. The calculation of the entropic contribution in binding energy is computationally expensive for a larger biomolecular complexes and hence it is omitted as similar to previous studies [21, 22, 70–72].

3. Results and discussion

In this chapter we employ sequence analysis, homology modeling, MD simulations, and binding energy calculation to (i) gain structural insights to the detailed binding mode, (ii) study atomic level tubulin isotypes-tauR2 interactions and (iii) study relative binding affinity between neuronal specific tubulin isotypes and TauR2.

3.1 Sequence analysis and homology modeling of neuronal specific tubulin isotypes

The residue composition of different β-tubulin isotypes mostly varies at the carboxy-terminal tail region as revealed by the multiple sequence alignment (**Figure 2**). The βI and βIII tubulin isotypes have longer C-terminal tail regions when compared with the βIIb tubulin isotype. The β-tubulin sequence in the template structure i.e., 6CVN (chain A) and human βIIb tubulin isotypes show 98.65% sequence identity. These sequence variations in the tubulin isotypes are reported to regulate number of protofilaments in the MT and their stability [73]. These β-tubulin isotypes sequences were used to generate three-dimensional homology models using 6CVN as the template. The structures of βI, βIIb, βIII tubulin isotypes were modeled using Modeler 9v20 [38]. The best homology model generated is selected using DOPE score. The DOPE score value for A and C chain of (i) βI subunits are −54299.89 and − 54291.24 (ii) βIIb subunits are −53487.13 and − 53054.42 and (iii) βIII subunits are −53725.86 and − 53054.42. The quality of these models is accessed using GMQE score [74], Verify3D [40], Errat score [41], Z-score [75] and Ramachandran plot [76, 77]. The parameters describing the overall quality of the modeled neuronal specific β-tubulin subunits are shorn in **Table 2**. The GMQE score provides an estimate of the accuracy of the modeled tertiary structure of neuronal specific β-tubulins. Here, GMQE score for all the modeled β-subunits is 0.98, which represents the accuracy of the generated model. Further, verify3D and ERRAT score also validates the quality of the generated models (**Table 1**). Ramachandran plots for all the modeled β-tubulin isotypes represents more than 98% of the residues occupy a favored region. The occupancy of amino acid residues in the Ramachandran plot is given in **Table 3**. These modeled structures of neuronal specific β-tubulin isotypes were used further to build the tubulin and TauR2 complexes such as βI/α/βI-TauR2, βIIb/α/βIIb-TauR2 and βIII/α/βIII-TauR2 using 6CVN.pdb as a template structure. These modeled complexes were used as starting structures to perform MD simulations.

3.2 Structural stability of the tubulin-TauR2 complexes

The all atom MD simulations were performed on tubulin-TauR2 complexes namely 6CVN-TauR2, 6CVN*-TauR2, βI/α/βI-TauR2, βIIb/α/βIIb-TauR2, βIII/α/βIII-TauR2 using Gromacs 2018.1 [78]. The stability of these tubulin-TauR2 complexes is accessed by plotting the potential energy during the simulation period, which highlight that all the complexes are well minimized, and simulation trajectories are well converged during the simulation period of 100 ns (**Figure 3**).

The parameters describing the stability of tau-tubulin complex such as RMSD (root mean square deviation), RMSF (root mean square fluctuation), and R*g* (radius of gyration) was studied. The RMSD values for simulated tubulin-TauR2 complexes, tauR2 and backbone atoms of tubulin trimer without considering

```
CLUSTAL O(1.2.4) multiple sequence alignment

6CVN:A|PDBID|CHAIN|SEQUENCE      MREIVHIQAGQCGNQIGAKFWEVISDEHGIDPTGSYHGDSDLQLERINVYYNEAAGNKYV   60
sp|Q9H4B7|TBB1_HUMAN             MREIVHIQIGQCGNQIGAKFWEMIGEEHGIDLAGSDRGASALQLERISVYYNEAYGRKYV   60
sp|Q9BVA1|TBB2B_HUMAN            MREIVHIQAGQCGNQIGAKFWEVISDEHGIDPTGSYHGDSDLQLERINVYYNEATGNKYV   60
sp|Q13509|TBB3_HUMAN             MREIVHIQAGQCGNQIGAKFWEVISDEHGIDPSGNYVGDSDLQLERISVYYNEASSHKYV   60
                                 ******** *************:*.:***** :*.  * * ******.****** ..***

6CVN:A|PDBID|CHAIN|SEQUENCE      PRAILVDLEPGTMDSVRSGPFGQIFRPDNFVFGQSGAGNNWAKGHYTEGAELVDSVLDVV   120
sp|Q9H4B7|TBB1_HUMAN             PRAVLVDLEPGTMDSIRSSKLGALFQPDSFVHGNSGAGNNWAKGHYTEGAELIENVLEVV   120
sp|Q9BVA1|TBB2B_HUMAN            PRAILVDLEPGTMDSVRSGPFGQIFRPDNFVFGQSGAGNNWAKGHYTEGAELVDSVLDVV   120
sp|Q13509|TBB3_HUMAN             PRAILVDLEPGTMDSVRSGAFGHLFRPDNFIFGQSGAGNNWAKGHYTEGAELVDSVLDVV   120
                                 ***:***********:**. :* :*:**.*:.*:******************::.**:**

6CVN:A|PDBID|CHAIN|SEQUENCE      RKESESCDCLQGFQLTHSLGGGTGSGMGTLLISKIREEYPDRIMNTFSVVPSPKVSDTVV   180
sp|Q9H4B7|TBB1_HUMAN             RHESESCDCLQGFQIVHSLGGGTGSGMGTLLMNKIREEYPDRIMNSFSVMPSPKVSDTVV   180
sp|Q9BVA1|TBB2B_HUMAN            RKESESCDCLQGFQLTHSLGGGTGSGMGTLLISKIREEYPDRIMNTFSVMPSPKVSDTVV   180
sp|Q13509|TBB3_HUMAN             RKECENCDCLQGFQLTHSLGGGTGSGMGTLLISKVREEYPDRIMNTFSVVPSPKVSDTVV   180
                                 *:*.*.********:.***************:.*:**********:***:**********

6CVN:A|PDBID|CHAIN|SEQUENCE      EPYNATLSVHQLVENTDETYCIDNEALYDICFRTLKLTTPTYGDLNHLVSATMSGVTTCL   240
sp|Q9H4B7|TBB1_HUMAN             EPYNAVLSIHQLIENADACFCIDNEALYDICFRTLKLTTPTYGDLNHLVSLTMSGITTSL   240
sp|Q9BVA1|TBB2B_HUMAN            EPYNATLSVHQLVENTDETYCIDNEALYDICFRTLKLTTPTYGDLNHLVSATMSGVTTCL   240
sp|Q13509|TBB3_HUMAN             EPYNATLSIHQLVENTDETYCIDNEALYDICFRTLKLATPTYGDLNHLVSATMSGVTTSL   240
                                 *****.**:***:**:*  :****************:************ ****:**.*

6CVN:A|PDBID|CHAIN|SEQUENCE      RFPGQLNADLRKLAVNMVPFPRLHFFMPGFAPLTSRGSQQYRALTVPELTQQMFDAKNMM   300
sp|Q9H4B7|TBB1_HUMAN             RFPGQLNADLRKLAVNMVPFPRLHFFMPGFAPLTAQGSQQYRALSVAELTQQMFDARNTM   300
sp|Q9BVA1|TBB2B_HUMAN            RFPGQLNADLRKLAVNMVPFPRLHFFMPGFAPLTSRGSQQYRALTVPELTQQMFDSKNMM   300
sp|Q13509|TBB3_HUMAN             RFPGQLNADLRKLAVNMVPFPRLHFFMPGFAPLTARGSQQYRALTVPELTQQMFDAKNMM   300
                                 **********************************::********:* ********::* *

6CVN:A|PDBID|CHAIN|SEQUENCE      AACDPRHGRYLTVAAVFRGRMSMKEVDEQMLNVQNKNSSYFVEWIPNNVKTAVCDIPPRG   360
sp|Q9H4B7|TBB1_HUMAN             AACDLRRGRYLTVACIFRGKMSTKEVDQQLLSVQTRNSSCFVEWIPNNVKVAVCDIPPRG   360
sp|Q9BVA1|TBB2B_HUMAN            AACDPRHGRYLTVAAIFRGRMSMKEVDEQMLNVQNKNSSYFVEWIPNNVKTAVCDIPPRG   360
sp|Q13509|TBB3_HUMAN             AACDPRHGRYLTVATVFRGRMSMKEVDEQMLAIQSKNSSYFVEWIPNNVKVAVCDIPPRG   360
                                 **** *:******* :***:** ****:*:*:* :*.:*** **********.*********

6CVN:A|PDBID|CHAIN|SEQUENCE      LKMSATFIGNSTAIQELFKRISEQFTAMFRRKAFLHWYTGEGMDEMEFTEAESNMNDLVS   420
sp|Q9H4B7|TBB1_HUMAN             LSMAATFIGNNTAIQEIFNRVSEHFSAMFKRKAFVHWYTSEGMDINEFGEAENNIHDLVS   420
sp|Q9BVA1|TBB2B_HUMAN            LKMSATFIGNSTAIQELFKRISEQFTAMFRRKAFLHWYTGEGMDEMEFTEAESNMNDLVS   420
sp|Q13509|TBB3_HUMAN             LKMSSTFIGNSTAIQELFKRISEQFTAMFRRKAFLHWYTGEGMDEMEFTEAESNMNDLVS   420
                                 *.*::*****.*****:*:*:**:*:***:****:****.****  ** ***.*::****

6CVN:A|PDBID|CHAIN|SEQUENCE      EYQQYQDATADEQG--EFEEEGEEDEA-----   445
sp|Q9H4B7|TBB1_HUMAN             EYQQFQDAKAVLEEDEEVTEEAEMEPEDKGH-   451
sp|Q9BVA1|TBB2B_HUMAN            EYQQYQDATADEQG--EFEEEEGEDEA-----   445
sp|Q13509|TBB3_HUMAN             EYQQYQDATAEEEG--EMYEDDEEESEAQGPK   450
                                 ****:***.*  :   *. *:   :
```

Figure 2.
Multiple sequence analysis of different β-tubulin isotypes. The βI, βIIb, βIII tubulin isotypes and template 6CVN show maximum residue variations mainly at C-terminal tail region. The TauR2 binding regions H12 helix and C-terminal tail region of β-tubulin subunits are shown in hot pink and brown, respectively.

S. No.	Chains	GMQE	Verify3D	Errat	z-score
1	β1 (A) -subunit	0.98	93.13%	81.3212	−8.93
2	β1 (C) -subunit	0.98	92.02%	83.2569	−8.78
3	β2b (A) -subunit	0.98	98.43%	87.471	−8.65
4	β2b (C) -subunit	0.98	98.43%	83.2947	−8.37
5	β3 (A) -subunit	0.98	98.44%	86.3636	−8.54
6	β3 (C) -subunit	0.98	92.67%	83.33	−8.39

Table 2.
Validation of three-dimensional models generated for βI, βIIb and βIII isotypes chain A and chain C using Swiss model GMQE score, Verify-3D, Errat score.

disordered C-tail were plotted over the trajectory. This analysis reveals the stability of all the studied complexes throughout the simulation time i. e. 100 ns. **Figure 4A** and **B** shows the RMSD plot for studied tubulin-TauR2 complexes and TauR2, respectively. The RMSD for the complex βIII/α/βIII-TauR2 is observed to be relatively more stable than other tubulin-TauR2 complexes. Similarly, structure

Region	β1 tubulin		β2 tubulin		β3 tubulin	
	Chain A	Chain C	Chain A	Chain C	Chain A	Chain C
% of most favored regions	98.4 (442)	98.9 (444)	98.9 (443)	98.9 (438)	98.9 (443)	98.9 (443)
% of additional allowed regions	1.3 (6)	0.9 (4)	0.7 (3)	1.1 (5)	1.1 (5)	0.7 (3)
% of outlier	0.2 (1)	0.2 (1)	0.4 (2)	0 (0)	0 (0)	0.4 (2)

Table 3.
Ramachandran plot showing the percentage of residues in the different regions for tubulin isotypes obtained from the Ramachandran plot using PROCHECK.

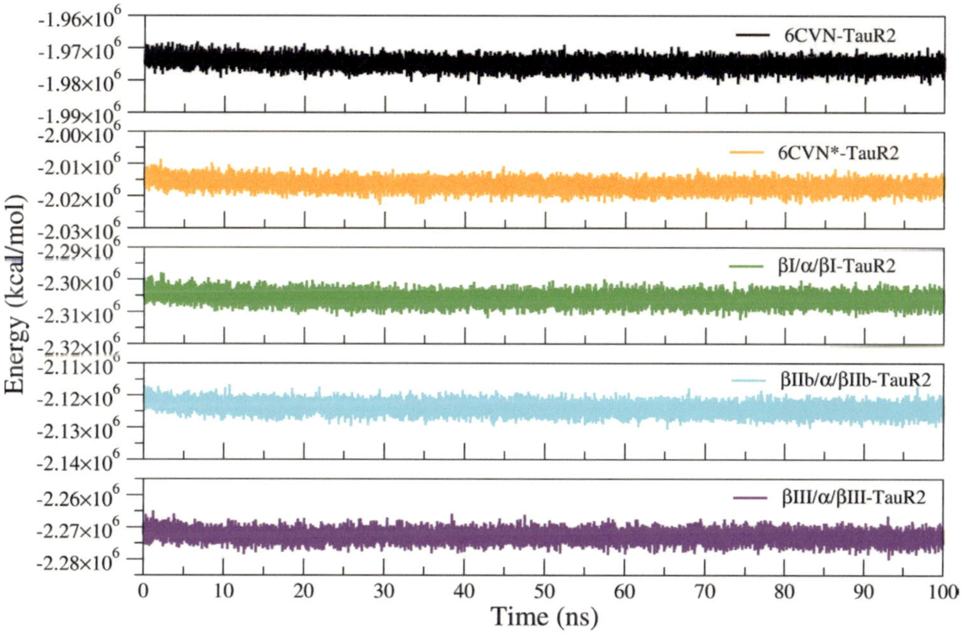

Figure 3.
Energy Minimization Plot Potential energy over the simulation time plotted for 6CVN-TauR2 (black), 6CVN-TauR2 (orange), βI/α/βI-TauR2 (green), βIIb/α/βIIb-TauR2 (cyan), βIII/α/βIII-TauR2 (violet) are shown.*

of TauR2 bound to βIII/α/βIII tubulin trimer expresses stable dynamics during the simulation. The complex 6CVN-TauR2 is stabilized at higher RMSD values, the primary reason for this elevated RMSD value is absence of C-tail region which highlights the importance of C-terminal tail in the stabilizing tubulin-TauR2 complex. Average backbone RMSD value is converged at ~3.5 Å hence represents the equilibration of all above simulated systems (**Figure 5**). The molecular dynamics simulation movies reveals the stable dynamics of all the simulated systems 6CVN-TauR2, 6CVN*-TauR2, βI/α/βI-TauR2, βIIb/α/βIIb-TauR2 and βIII/α/βIII-TauR2 (https://youtu.be/mU2Jrm5jusY, https://youtu.be/Sr2JiQWha9A, https://youtu.be/U5S6X-o8kO8, https://youtu.be/xYbm9eCsE4Q, https://youtu.be/0H0CsmveT24) respectively. Further, specificity of TauR2 towards tubulin subunits was accessed by replacing the TauR2 with negative control 'polyA' peptide of same length. Interestingly, This system having negative control poly A bound to 6CVN* shows the weak binding during the simulation. These weaker interactions of polyA peptide with tubulin subunits (https://youtu.be/ZEFQblQTHqk) represents that tauR2 has specificity towards tubulin subunits. s.

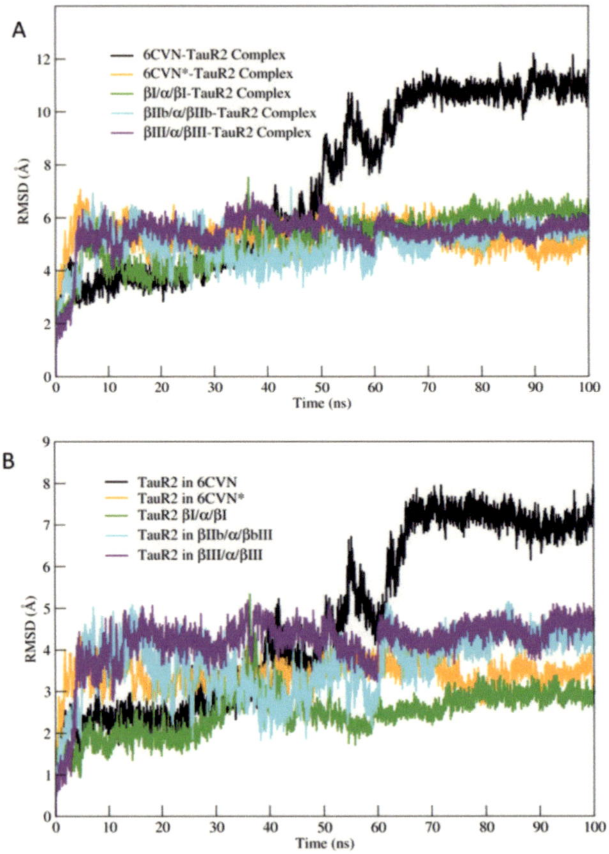

Figure 4.
Stability of the tubulin-TauR2 complex and TauR2. (A) The Root mean square deviation values (RMSD) for tubulin-tauR2 complexes. RMSD values for 6CVN, 6CVN, βI/α/βI, βIIb/α/βIIb and βIII/α/βIII have been plotted in black, orange, green, cyan and violet, respectively. (B) The Root mean square deviation values for TauR2 shown using same color Scheme as in (A).*

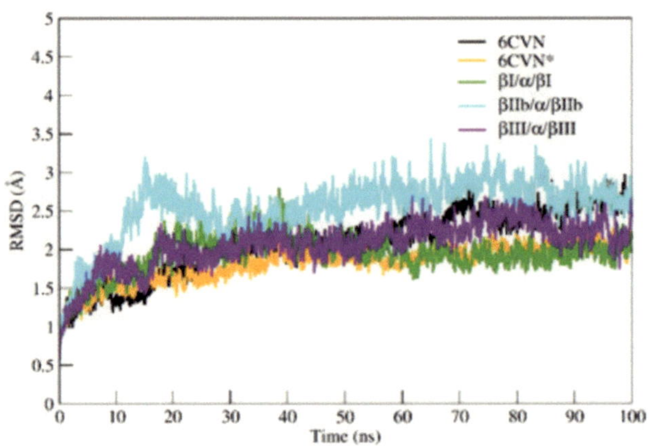

Figure 5.
Backbone Root mean square deviation for different tubulin subunits. Color scheme is same as ***Figure 3****.*

3.3 Residue fluctuations of tubulin subunits and TauR2

The flexibility of tubulin trimers systems and TauR2 has been studied using RMSF analysis. For this analysis Cα-atom from the backbone was selected to get fluctuations in the overall protein. **Figure 6** represents the RMSF for tubulin subunits and TauR2. The residues from the H12 helix of β-tubulin and the C-terminal tail region (residue 400–451) show significant decrease in the RMSF values, as their free dynamics is arrested by tau binding (**Figure 6A** and **B**). RMSF

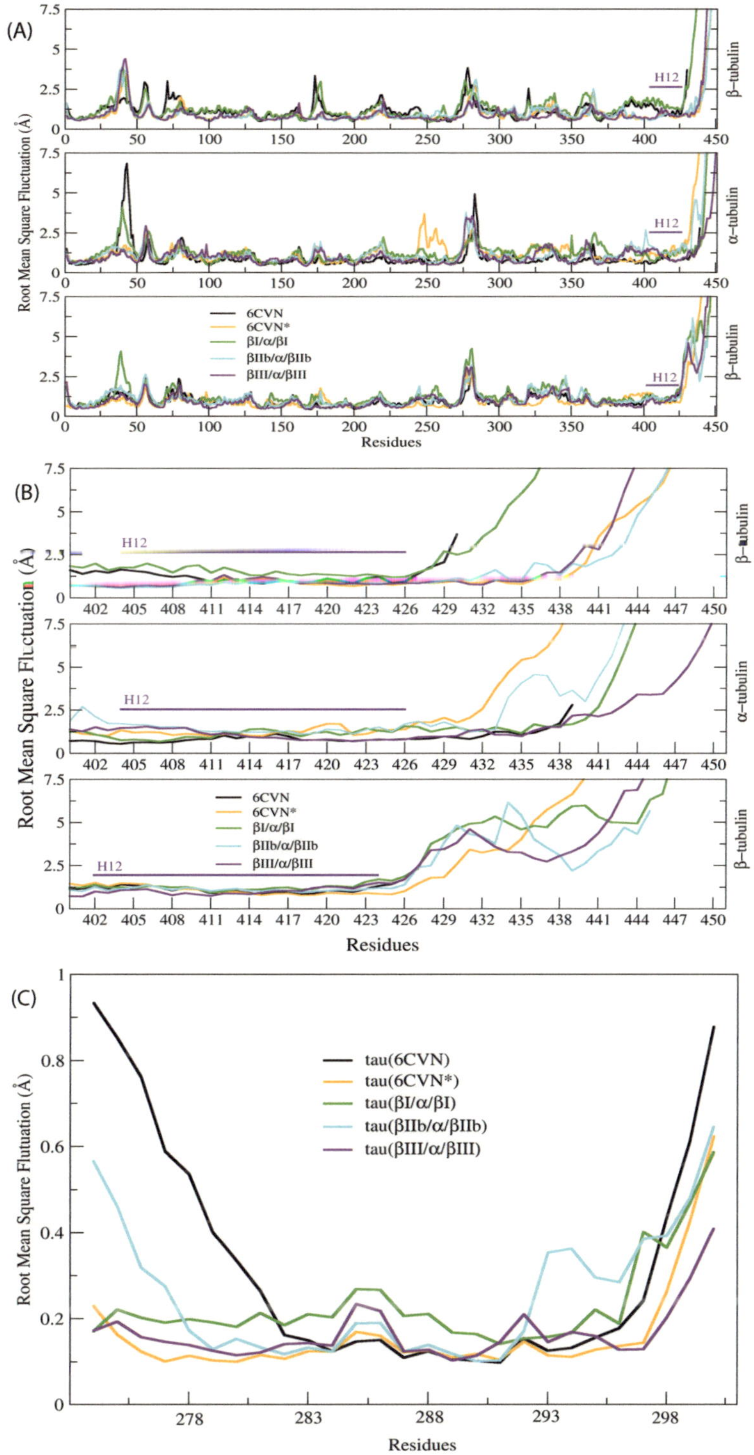

Figure 6.
Root mean square fluctuations (RMSF) of different β/α/β tubulin subunits and TauR2 (A) RMSF of different β/α/β tubulin subunits (B) Magnified view of their C-terminal H12 helix and tail regions (C) RMSF of TauR2 bound with different β/α/β tubulin subunits observed during the simulations5. Color scheme is same as ***Figure 3****.*

values for the tubulin β-subunits in the systems 6CVN*, βIIb/α/βIIb and βIII/α/βIII are lesser than those of 6CVN and βI/α/βI tubulin subunits (**Figure 6B**). This observation also highlights the binding of TauR2 at the interdimer interface where residual fluctuations are less. However, the part of C-tail region which has no direct contact with TauR2 is highly flexible (**Figure 6B**). The H12 helix and C-terminal tail region of the tubulin subunits significantly contribute to the non-covalent interactions resulting towards stronger binding of TauR2. Therefore, these intermolecular interactions were analyzed in detail and are discussed in the section '*Intermolecular interactions between tubulin and tau*'. Further, atomic Cα-fluctuations of TauR2 (**Figure 6C**) was also studied for better understanding its conformational behavior during the MD simulations. It is surprising to observe highest fluctuations at the N- and C-terminal region in TauR2 bound to 6CVN, where the C-terminal tail region is absent (**Figure 6C**). Interestingly, residual fluctuations expressed by TauR2 bound to βIII/α/βIII-tau complex are much lesser as compared to 6CVN*-TauR2, βI/α/βI-TauR2 and βIIb/α/βIIb-TauR2 complexes (**Figure 6C**). This also proves that the C-terminal tail region of tubulin subunits plays an important role in the binding of TauR2.

Overall, RMSF analysis suggests the significance of H12-helix and C-terminal tail region in stabilization of the microtubule by binding of tau repeats (TauR2) and it also reveals the greater affinity of TauR2 towards βIII tubulin isotypes which are overexpressed in neuronal cells and brain. Further compactness of all the tubulin-TauR2 complexes was explored by calculating the radius of gyration (R_g) and this analysis is discussed in the next section.

3.4 Compactness of tubulin-TauR2 complexes

The radius of gyration (R_g) indicates the level of compactness of the protein system which is helpful in getting an insight into the stability of the protein–protein complex. It also helps to understand folding or unfolding of protein structure during the simulation. The R_g values for all the studied tubulin-TauR2 complex ranges from 38.8–40.5 Å (**Figure 7A**). The complex βIII/α/βIII-TauR2 shows stable R_g value for the entire simulation period however other complexes 6CVN-TauR2 6CVN*-TauR2, βI/α/βI-TauR2, βIIb/α/βIIb-TauR2 show variations in their R_g values. The absence of C-terminal tail region in the complex 6CVN-TauR2 leads to the less R_g values when compared to other tubulin-TauR2 (**Figure 7A**). **Figure 7B** represents the R_g values of only TauR2 in different tubulin-TauR2 complexes. The R_g values for TauR2 shows fluctuations between 17.5 to 20 Å in case of 6CVN*, βIIb/α/βIIb, and βIII/α/βIII complexes except for βI/α/βI complex (**Figure 7B**). The βIII tubulin subunits show R_g value of ~18 Å and βI tubulin subunits have largest R_g value of 22.5 Å as shown in **Figure 7B**. On the other hand, TauR2 bound to 6CVN shows uninterrupted decline in R_g values from 21.5 Å to 16.5 Å. This analysis also highlights the importance of C-terminal tail region in the stable binding of tau (**Figure 7B**). It is important to note that βIII tubulin subunits (**Figure 8**) have R_g values like that of βIII/α/βIII-TauR2 complex (**Figure 7A**). This highlights that the tubulin subunits composed of βIII tubulin isotype are structurally stable after binding to the TauR2. Thus, calculation of R_g values for tubulin-TauR2 complexes, tubulin subunits and TauR2 reveals (i) structural stability of the βIII/α/βIII-tau complex over other complexes and (ii) importance of the C-terminal tail region in the binding of TauR2. Contact surface area (CSA) and solvent accessible surface area (SASA)

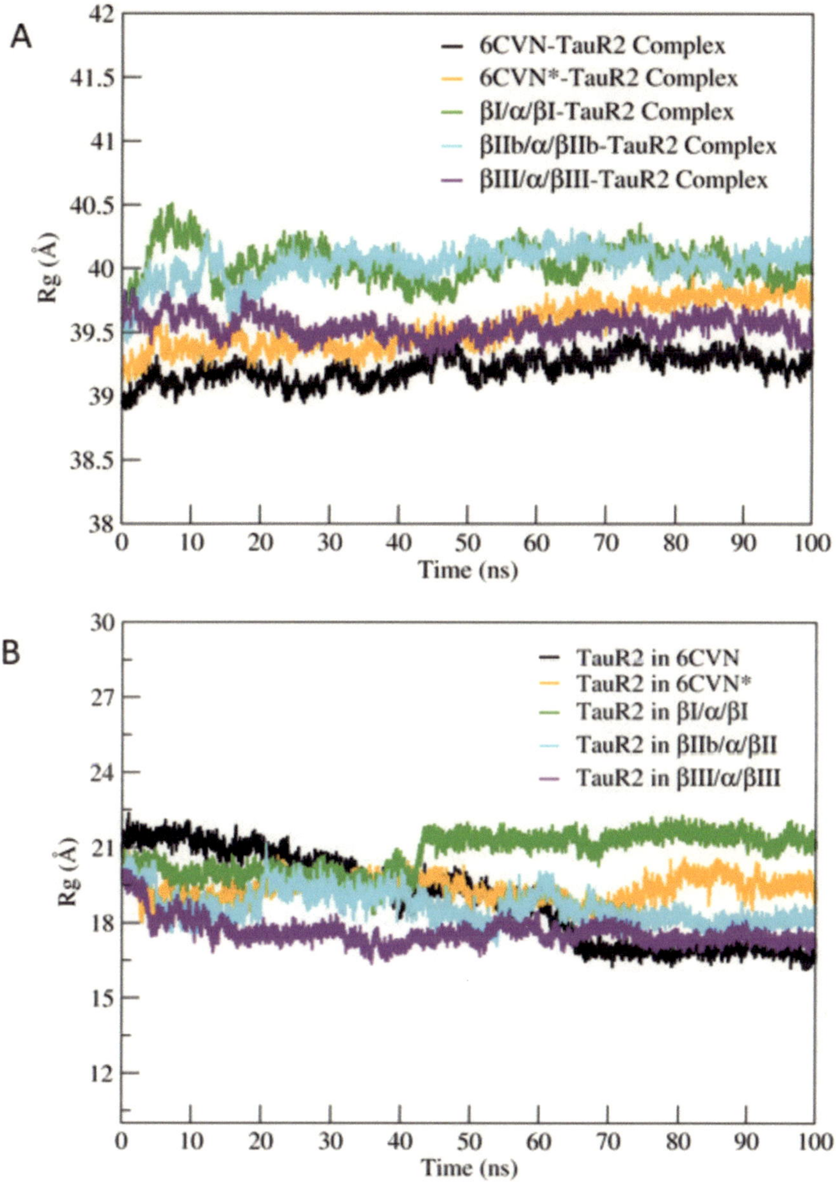

Figure 7.
Radius of Gyration (Rg) of different tubulin-TauR2 complexes and TauR2. (A) Rg of 6CVN-TauR2 (black), 6CVN-TauR2 (orange), βI/α/βI-TauR2 (green), βIIb/α/βIIb-TauR2 (cyan), βIII/α/βIII-TauR2 (violet) (B) Rg for TauR2 in different tubulin-TauR2 complexes. Color scheme same as* **Figure 3**.

was calculated using 'gmx sasa' tool of gromacs to understand the exposure of the interface residues of tubulin subunits bound to the TauR2 [46].

3.5 Contact surface area (CSA) and solvent accessible surface area for tubulin-TauR2 complexes

The CSA and SASA describes the accessibility of a binding interface and protein surface to the solvent, respectively. It is well documented that, TauR2 binds to the MT exterior surface via C-terminal tail region [8, 79–82]. Therefore,

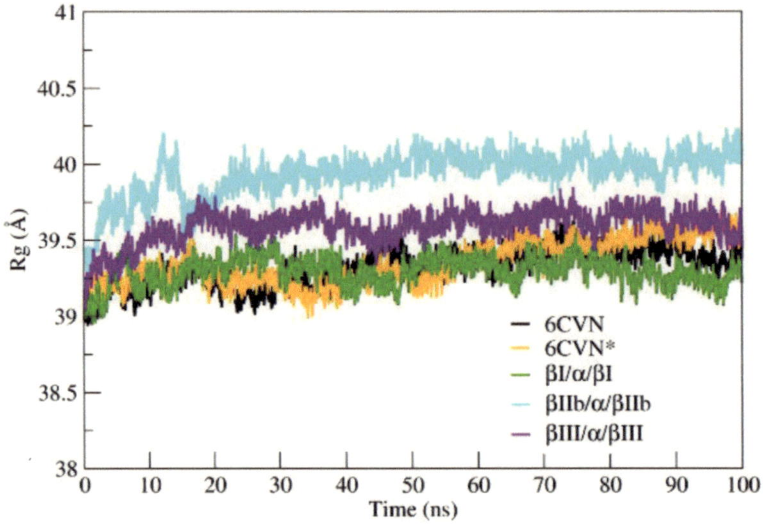

Figure 8.
Radius of Gyration for different tubulin isotypes. Color scheme is same as ***Figure 3****.*

initially contact surface area (CSA) of the interface between the TauR2 and tubulin trimer, was calculated, without considering flexible C-terminal tail region. The CSA of βIII/α/βIII is very less when compared to other tubulin isotypes (**Figure 9A**) this represents the tight binding of TauR2 to the βIII/α/βIII tubulin subunits. The higher CSA for βI/α/βI-TauR2 complex indicates weaker binding of the TauR2 to the βI/α/βI tubulin subunits. Furthermore, least SASA in complex βIII/α/βIII-TauR2 represents tight binding of TauR2 to the βIII/α/βIII (**Figure 9B**). On the other hand, higher hydrophobic SASA of the complex βI/α/βI-TauR2 indicate the exposure of hydrophobic residues which are responsible for loss of native contacts between tubulin and TauR2. The SASA for 6CVN*, βI/α/βI, βIIb/α/βIIb, βIII/α/βIII shows higher SASA values between 4900 and 5400 Å when compared to 6CVN-TauR2 (~4500 Å) due to the presence of C-terminal tail region (**Figure 10**). To get detailed understanding of the atomic-level interaction between tubulin isotypes and TauR2, further hydrogen bonding interactions were estimated during simulation and in the MD simulated end-structures obtained from trajectory.

3.6 Intermolecular interactions between tubulin and TauR2 in tubulin-TauR2 complexes

The total number of hydrogen bonds formed between tubulin isotypes and TauR2 during the MD simulations are calculated using in-built *'gmx hbond'* command [46]. The cut-off value for the measurement of H-bond was set to 3.4 Å. Consistent H-bond formation was observed throughout the MD simulation in all tubulin-TauR2 complexes. The average number of H-bonds roughly varies between 10 to 20 as shown in **Figure 11**. The details of atom participating in the hydrogen bonding interactions present between tubulin isotypes and TauR2 in the MD simulation end-structures are listed in **Table 4**. All the hydrophobic interactions participating in the formation of stable tubulin-TauR2 complexes are listed in **Table 5**. The βIII/α/βIII-TauR2 complex shows the maximum number of electrostatic interactions when compared to other tubulin-TauR2 complexes (**Table 6**). Further, to understand the role of TauR2

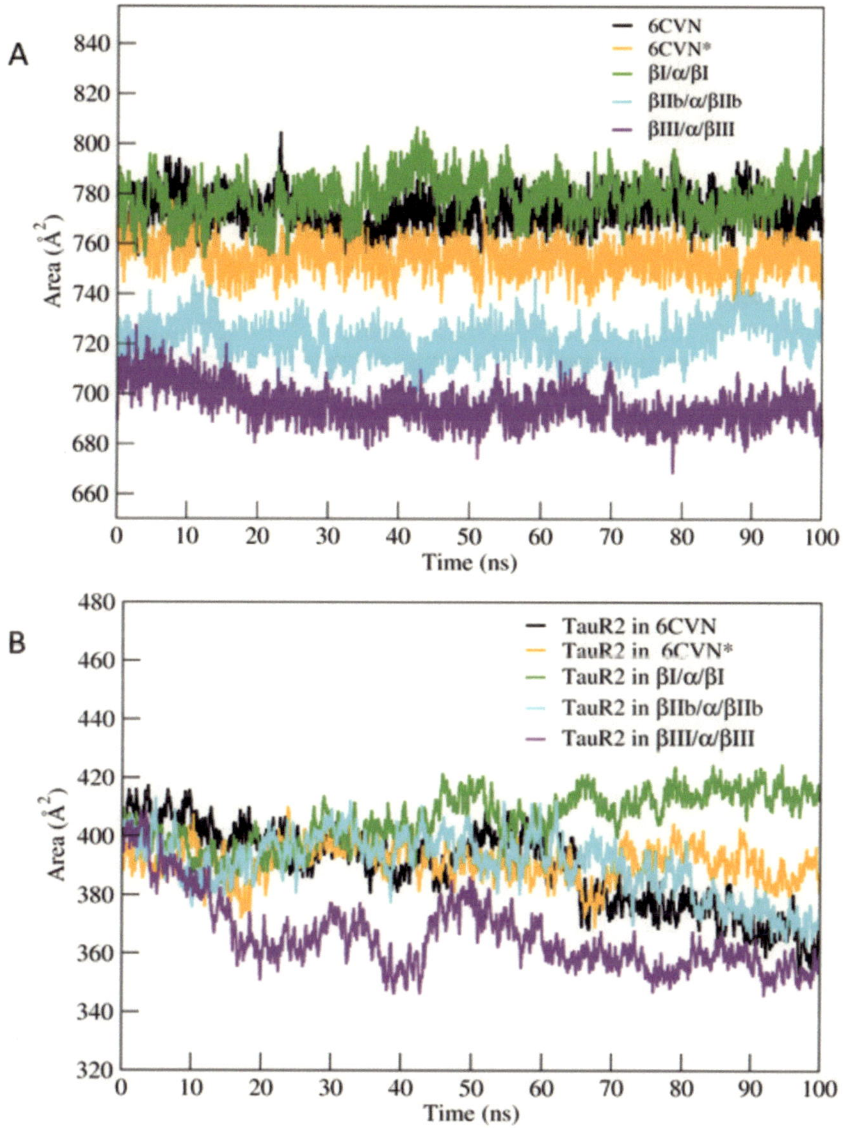

Figure 9.
Contact surface area (CSA) and solvent accessible surface area (SASA) of different β/α/β-tubulin subunits andTauR2. (A) CSA for different 6CVN-TauR2 (black), 6CVN-TauR2 (orange), βI/α/βI-TauR2 (green), βIIb/α/βIIb-TauR2 (cyan), βIII/α/βIII-TauR2 (violet) complexes. (B) hydrophobic SASA for tubulin isotype bound TauR2. Color scheme same as* ***Figure 3****.*

in stabilizing tubulin subunits, secondary structure analysis on TauR2 was done using DSSP.

3.7 Conformational changes in TauR2 upon tubulin binding

Tau belongs to the class of intrinsically disordered proteins for which no definitive secondary structure exists. Hence their structure determination is difficult by using existing biophysical techniques like X-ray crystallography and NMR. Previous experimental observations propose that tau repeat undergoes a conformational changes from the disordered to ordered state when it binds to the MT [2, 83–86]. Hence, the secondary structural changes during the MD simulations in the TauR2

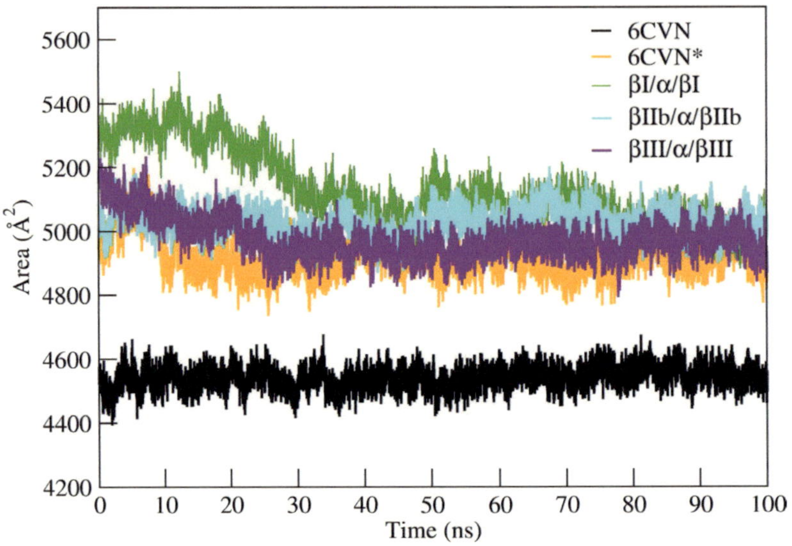

Figure 10.
Solvent accessible surface area for different tubulin subunits. SASA plotted for 6CVN (black), 6CVN (orange), βI/α/βI (green), βIIb/α/βIIb (cyan), βIII/α/βIII (violet) are shown.*

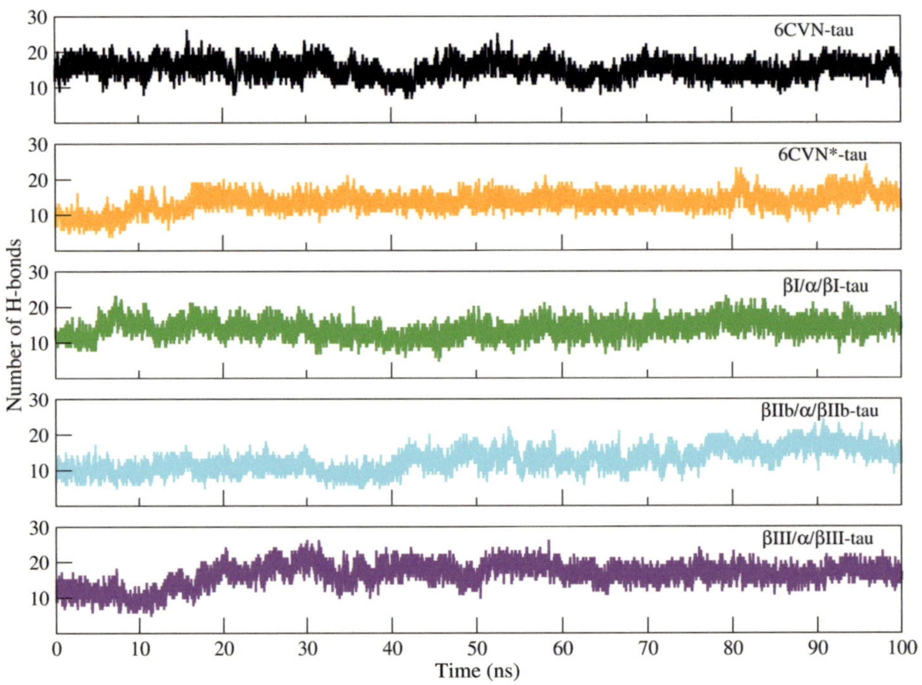

Figure 11.
The number of hydrogen bonds formed in between tubulin subunits and TauR2 during MD simulation. Color scheme is same as **Figure 3**.

were studied using DSSP [56]. **Figure 12** represents conformational changes in the secondary structure of TauR2 upon binding to the tubulin. TauR2 in 6CVN- TauR2 (**Figure 12A**) and 6CVN*- TauR2 complexes (**Figure 12B**) show formation of short and transient 3_{10}-helix during the simulation. The TauR2 in βI/α/βI- TauR2 complex does not form either α-helix or transient 3_{10}-helix as shown in **Figure 12C**. The TauR2 in βIIb/α/βIIb-TauR2 complex shows the formation of short-lived α-helix and

System	Atoms involved in H-bonding	Distance (Å)	Angle (°)
6CVN-TauR2	D: SER16: HG - B: GLU434:OE2	1.55968	170.912
	C: LYS392:HZ2 - D: ASP22:OD1	1.79981	155.003
	D: SER20:H - B: GLU434:OE1	1.80025	149.621
	D: CYS18:H - B: ASP431:OD1	1.8071	147.422
	D: GLY19:H - B: ASP431:OD1	1.81516	170.051
	D: ASN23:HD21 - C: PHE389:O	1.83614	168.422
	D: ILE5:H - B: GLU415:OE1	1.87843	148.138
	B: ARG402:HH22 - D: LYS7:O	1.95256	131.998
	D: LYS21:HZ3 - B:ASP438:O	1.96835	128.744
	C: ARG391: HE - D: SER20:O	1.97398	162.451
	D: ASN6:H - B: GLU415:OE2	1.97803	165.667
	D: LYS17:HZ2 - B: ASP424:OD1	2.01656	167.691
6CVN*-TauR2	A: SER16: HG - E: ASP431:OD2	1.56368	161.89
	A: ASN6:HD21 - G: GLN433:OE1	1.68122	165.335
	A: LYS1:HZ1 - G: ASP417:OD1	1.73491	152.719
	A: LYS7:HN - E: ALA400:O	1.75148	168.412
	A: SER12: HG - A: ASP10:OD2	1.77457	163.204
	E: LYS401:HZ1 - A: ASN6:OD1	1.77792	156.289
	A: SER20:HN - E: GLU434:O	1.8254	168.05
	A: ILE4:HN - G: GLN424:OE1	1.85744	159.004
	F: ARG391:HH21 - A: SER20: OG	1.88141	150.648
	A: LYS17:HN - E: ASP431:OD2	1.88237	156.579
	A: LYS1:HT2 - G: ASP417:OD2	1.93321	163.221
	A: LYS25:HZ2 - A: VAL27: OXT	1.95566	155.991
	F: ARG391: HE - A: SER20:O	1.95954	137.171
	A: SER12:HN - A: ASP10:OD2	1.97395	166.966
	A: VAL2:HN - G: GLU421:OE2	1.97952	167.491
	A: SER20: HG - E: TYR262: OH	2.01579	158.876
	A: CYS18:HN - E: ASP431:OD2	2.04985	143.832
	A:ASP22:HN - A: ASP22:OD1	2.06458	123.428
βI/α/βI-TauR2	D: SER20: HG - B: GLU434:OE2	1.71015	154.981
	D: LYS17:H - B: ASP431:OD2	1.73879	154.124
	D: SER16: HG - B: ASP431:OD2	1.75545	159.778
	D: VAL2:H - A: GLU421:OE1	1.79605	175.632
	D: SER20:H - B: GLU434:OE2	1.79906	165.806
	D: LYS21:HZ1 - B: GLU434:O	1.82861	143.493
	B: LYS430:HZ2 - D: VAL14:O	1.84508	147.656
	D: ASN23:HD22 - C: PHE389:O	1.87483	147.025
	D: LYS1:H3 - A: ASP417:OD1	1.96013	159.665
	D: LYS7:H - B: ALA400:O	2.03926	154.297
	D: ILE4:H - A: GLN424:OE1	2.05966	141.181

System	Atoms involved in H-bonding	Distance (Å)	Angle (°)
βIIb/α/βIIb-TauR2	A: SER16: HG - E: ASP431:OD1	1.71023	174.429
	A: LYS17:HZ2 - E: ASP424:OD2	1.72125	162.879
	A: LYS1:HT2 - F: GLU421:OE2	1.73767	160.093
	A: LYS25:HZ2 - E: GLU445:OE1	1.7524	156.386
	A: LYS7:HN - E: ALA400:O	1.7726	158.577
	A: ASN6:HD22 - F: ASP431:OD1	1.88399	166.501
	A: LYS17:HN - E: ASP431:OD1	1.90489	145.148
	E: ARG402:HH12 - A: LYS7:O	1.90591	147.259
	A: VAL27:HN - E: GLU445:OE1	1.92907	160.86
	A: CYS18:HN - E: ASP431:OD1	2.03765	145.506
	A: LYS25:HZ3 - E: GLU446:O	2.04646	165.654
	F: GLN424:HE21 - A: LYS1:O	2.06218	173.384
βIII/α/βIII-TauR2	A: SER16: HG - E: ASP431:OD1	1.57372	172.137
	A: LYS21:HZ1 - E: GLU443:OE2	1.74362	173.969
	F: GLN424:HE21 - A: VAL2:O	1.79549	177.625
	A: GLN15:HE21 - E: GLU443:OE2	1.81621	154.684
	A: LYS8:HZ3 - F: GLU433:OE1	1.84962	164.797
	A: LYS8:HZ1 - E: ASP396:OD1	1.85984	168.252
	G: ARG391:HH11 - A: ASN23:OD1	1.89423	162.494
	E: GLY442:HN - A: ILE24:O	1.93324	171.538
	A: LYS17:HN - E: ASP431:OD1	1.99492	142.974
	A: CYS18: HG - E: ASP431:OD1	2.0734	155.708

Table 4.
Hydrogen bonding interaction between tubulin subunits and TauR2 after molecular dynamics simulations.

System	Hydrophobic Interactions	Distance (Å)
6CVN-TauR2	B: ALA427 - D: LYS17	4.23972
	B: ARG264 - D: CYS18	4.32305
	B: ALA426 - D: VAL14	4.34456
	B: ARG402 - D: ILE4	5.12692
	B: VAL409 - D: ILE4	5.16258
	B: ARG422 - D: LEU11	5.2126
	B: ALA427 - D: VAL14	5.2724
6CVN*-TauR2	E: ALA426 - A: VAL14	3.78619
	E: ALA426 - A: LEU11	4.1983
	E: ALA400 - A: LYS8	4.29681
	G: PHE260 - A: ILE4	4.3943
	A: VAL2 - G:PRO261	4.85577
	A: LEU9 - A: LEU11	5.10635
	A: LYS21 - E: VAL437	5.11425
	E: ALA427 - A: VAL14	5.28218
	A: VAL14 - A: LEU11	5.3507
	E: ARG422 - A: LEU9	5.37331
	A: VAL2 - A: ILE4	5.42169

System	Hydrophobic Interactions	Distance (Å)
βI/α/βI-TauR2	B: ALA426 - D: LEU11	4.086
	B: ALA426 - D: VAL14	4.25902
	A: PHE425 - D: ILE4	4.29071
	B: ALA427 - D: VAL14	4.57533
	A: PHE260 - D: VAL2	4.62894
	B: TYR262 - D: CYS18	4.75533
	B:PRO263 - D: LYS17	4.89022
	B: ARG402 - D: LYS7	4.99921
	D: CYS18 - B: VAL435	5.09522
	B: ARG422 - D: LEU9	5.15149
	B: LYS430 - D: VAL14	5.38523
	A: ALA428 - D: ILE4	5.40848
	B: VAL440 - D: LYS21	5.41789
	C: ILE405 - D: ILE24	5.46457
βIIb/α/βIIb-TauR2	E: ALA427 - A: LYS17	3.91294
	E: ALA426 - A: VAL14	3.93471
	E: ALA426 - A: LEU11	4.19609
	A: CYS18 - E: ARG264	4.36259
	E: ALA400 - A: LYS8	4.4046
	A: CYS18 - E:PRO263	5.04073
	A: CYS18 - E: ILE265	5.21553
	G: LYS392 - A: ILE24	5.27213
	E: TYR399 - A: LEU9	5.32769
βIII/α/βIII-TauR2	E: ALA426 - A: LEU11	3.83867
	E: ALA426 - A: VAL14	4.14469
	E: ALA427 - A: LYS17	4.22472
	E: ALA427 - A: VAL14	4.83218
	E: ARG422 - A: LEU11	4.84097
	A: LYS25 - E: VAL437	5.06163
	A: CYS18 - E: ARG264	5.14951
	E: ARG422 - A: LEU9	5.48905

Table 5.
Hydrophobic interactions between different β/α/β-tubulin isotypes and TauR2 after molecular dynamics simulations.

3_{10}-helix (**Figure 12D**). The terminal residues of TauR2 from Ser293 to Val300 in βIII/α/βIII-TauR2 complex () undergoes turn to α-helix conformational transition (**Figure 12E**). Therefore, it is proposed that this conformational transition of TauR2 from disordered to ordered state promotes the stable binding of TauR2 with the βIII/α/βIII tubulin subunits.

3.8 Relative binding affinity of TauR2 towards neuronal specific tubulin isotypes

The relative binding affinity of TauR2 towards neuronal specific tubulin isotypes (β/α/β) was analyzed by performing MMPBSA calculations for complexes

Systems	Electrostatic interactions	Distance (Å)
6CVN-TauR2	D: LYS8:NZ - A: ALA430:O	4.04432
	D: LYS21:NZ - B: SER439:O	4.66847
	D: LYS25:NZ - B: GLU434:OE2	4.90999
	D: LYS21:NZ - B: GLU434:OE1	5.38615
6CVN*-TauR2	A: LYS1: N - G: GLU421:OE2	4.85022
	A: LYS7:NZ - E: GLU415:OE1	4.9846
	A: LYS25:NZ - E: GLU441:OE1	5.12197
	A: LYS17:NZ - E: ASP424:OD2	5.27557
βI/α/βI-TauR2	D: LYS1: N - A: GLU421:OE1	2.86182
	D: LYS25:NZ - C: GLU412:OE1	4.31321
	D: LYS7:NZ - B: GLU415:OE1	4.45715
	D: LYS21:NZ - B: GLU434:OE2	4.75044
βIIb/α/βIIb-TauR2	A: LYS21:NZ - E: GLU434:OE2	4.3529
	A: LYS8:NZ - E: ASP396:OD2	5.021
βIII/α/βIII-TauR2	A: LYS25:NZ - E: GLU434:OE2	2.68494
	A: LYS25:NZ - E: GLU450:OE2	2.85019
	A: LYS1: N - F: ASP417:OD2	2.91844
	A: LYS1: N - F: GLU421:OE2	4.28381
	A: LYS7:NZ - E: GLU415:OE1	4.31182
	A: LYS21:NZ - E: GLU434:OE1	4.48844
	A: LYS17:NZ - E: ASP424:OD2	4.70401
	G: LYS392:NZ - A: ASP22:OD2	4.95323
	A: LYS21:NZ - E: GLU450:OE2	5.37409

Table 6.
Electrostatic interactions between different β/α/β-tubulin isotypes and TauR2 after molecular dynamics simulations.

6CVN-tau, 6CVN*-tau, βI/α/βI-tau, βIIb/α/βIIb-tau and βIII/α/βIII-tau etc. The energy components that govern the binding energy are recorded in **Table 7**. This analysis reveals that, βIII/α/βIII-tau complex shows most favorable interactions while 6CVN-tau complex is least favorable as supported by the binding energy values listed in **Table 7**. Thus, it is interesting to note the significance of C-terminal tail of the tubulin subunits in the stable binding of the tau repeat R2 to stabilize this complex. The order of calculated binding energy in between TauR2 with neuronal specific tubulin-TauR2 complexes is βIII/α/βIII > βIIb/α/βIIb >6CVN* > βI/α/βI > 6CVN. The electrostatic interactions in these complexes contribute significantly to the binding energy particularly in βIII/α/βIII-TauR2 and βIIb/α/βIIb-TauR2 complexes when compared to the 6CVN and βI/α/βI tubulin subunits (**Table 7**). The complex βI/α/βI-TauR2 exhibits higher binding energy leading to its weaker affinity towards TauR2. In addition to βIII/α/βIII-TauR2 the complex βIIb/α/βIIb-TauR2 also exhibits relatively higher affinity towards TauR2 compared to rest other complexes. Further, contribution of the individual residues in the binding energy has been investigated by calculating the decomposition energy for each residue. This analysis reveals that, residues from the H12 helix and C-terminal tail of tubulin subunits shows maximum contribution in the binding energy (**Figure 13**). The per-residue

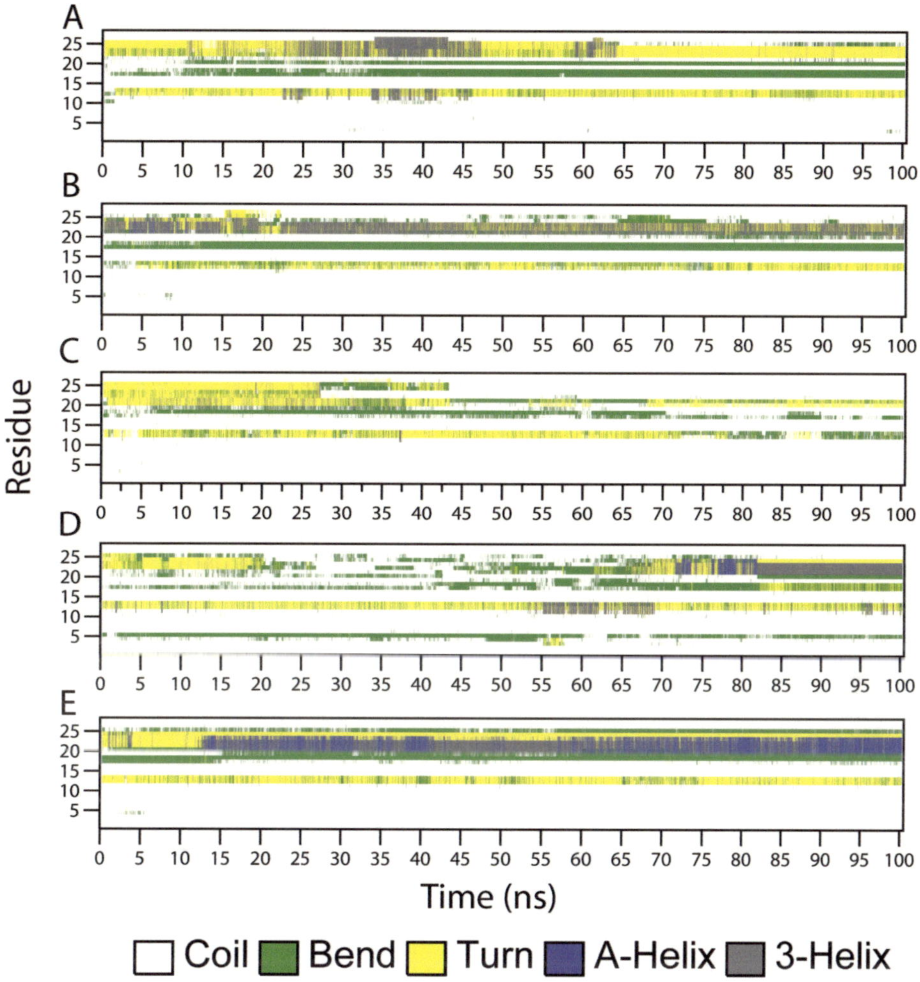

Figure 12.
The secondary structure changes during MD simulation using DSSP for TauR2. Secondary structure changes observed in (A) 6CVN-TauR2 (B) 6CVN-TauR2 (C) βI/α/βI-TauR2, (D) βIIb/α/βIIb-TauR2 and (E) βIII/α/βIII-TauR2.*

System	6CVN	6CVN*	βI/α/βI	βII/α/βII	βIII/α/βIII
Vdw	−97.25 ± 0.55	−131.17 ± 0.62	−133.38 ± 0.69	−124.36 ± 0.60	−125.24 ± 0.59
Elec	−1232.33 ± 3.43	−1534.36 ± 3.12	−1423.07 ± 2.77	−1659.30 ± 4.68	−1768.65 ± 2.77
Polar	413.98 ± 4.73	349.43 ± 4.18	439.27 ± 2.95	433.79 ± 4.13	505.14 ± 2.92
SASA	−12.35 ± 0.06	−15.12 ± 0.07	−17.05 ± 0.05	−15.66 ± 0.06	−16.04 ± 0.05
Binding Energy	−927.87 ± 3.15	−1331.15 ± 3.19	−1134.13 ± 1.13	−1365.2.26 ± 2.26	−1404.7 ± 1.84

Table 7.
The relative binding energy of the tubulin-TauR2 complexes calculated using MMPBSA. All energies are given in kcal/Mol.

interactions energy (residue decomposition energy) calculated for various pairs of interacting residues highlights the importance of the interacting residues. The residues from the H12-helices and C-terminal tail region of complex βIII/α/βIII shows maximum contribution (most negative energy) in the non-bonded contacts leading to the stable and tight binding of the tauR2 to the βIII/α/βIII tubulin subunits.

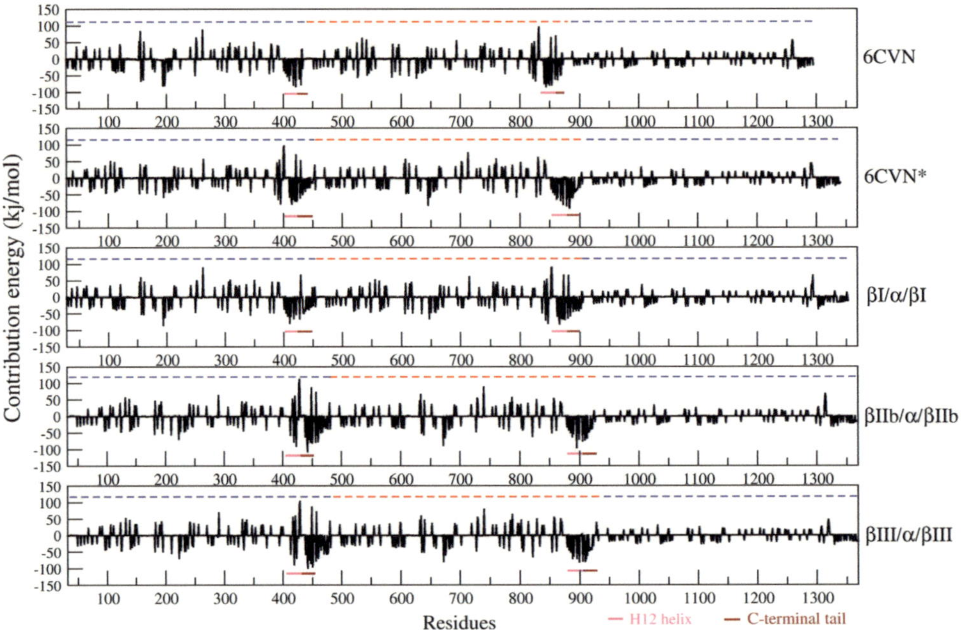

Figure 13.
The H12 and C-terminal tail regions show highest energy contribution for the binding of TauR2 in 6CVN, βI/α/βI, βIIb/α/βIIb and βIII/α/βIII tubulin subunits except in case of 6CVN which does not have C-terminal tail region.*

Hence, relative binding energy calculations further support all other MD simulation results highlighting the tight binding of TauR2 to the βIII/α/βIII tubulin isotype which is predominantly expressed in the neuronal cells and brain.

4. Conclusion

MTs are distributed across all types of cells and play an important role in the cellular functions. Structurally MTs are made up of α/β heterodimeric subunits. Large diversity of α and β-tubulin isotypes exists which are differently expressed in different types of cells, this makes MTs unique from one another in relative proportion of isotypes. The much elevated expression levels of βII and βIII tubulin isotypes about 58% and 25% respectively have been reported to in neuronal cells and brain [35]. The present study extensively uses molecular modeling approaches including homology modeling, MD simulation, binding energy to investigate the binding mode and interaction of neuronal specific tubulin isotypes with TauR2.

Extensive analysis on MD simulation trajectory shows a stable complex formation in between different tubulin isotype and TauR2. The stability of these complexes is mainly mediated by the interactions of H12 helix and C-terminal tail of the α/β tubulin isotypes with TauR2. TauR2 shows differential binding affinity towards various neuronal specific β-tubulin isotypes (βI, βII and βIII) the order of binding affinity is 'βIII> βIIb>βI'. Thus, it is found that TauR2 expresses greater binding affinity with βIII and βIIb tubulin isotypes which are abundantly expressed in neuronal cells and brain. The molecular modeling strategy adopted in this chapter could be potentially used to understand differential binding affinity of other tau repeats such as R1, R3, R4 towards β tubulin isotypes present in other cell lines. The structures for other repeats could be generated using homology modeling and their interactions with neuronal specific tubulin isotypes could also be studied using similar molecular modeling approach.

I believe that the knowledge on precise molecular origin of differential binding affinity of tau with β tubulin isotypes present different cell types will pave the way for developing effective treatments against tau related disorders such as Alzheimers, Amyotrophic lateral sclerosis and other tauopathy linked neurodegenerative disorders.

Thus, homology modeling allows us to investigate important biomolecular interactions whose protein structures are not known and/or difficult to get by using biophysical experiments. This, homology modeling, computational tool has made it easier to address various challenging problems in understanding the basic phenomenon's/pathways in modern biology. Also, it has proven wide successful applications in determining the role of proteins in various genetic diseases, hormonal disorder cancers, neurological disorders and other diseases etc. by exploring protein structure and functions using molecular modeling techniques.

Acknowledgements

VVB is thankful to IIT Bombay for Institute postdoctoral fellowship. Author is also thankful to Prof. Ambarish Kunwar, Department of Biosciences and Bioengineering, IIT Bombay, Mumbai for fruitful discussion and providing necessary computational resources to perform this research work. Author also sincerely thanks Creative Commons license for giving permission to use data from my manu script my published research article in Scientific Reports (https://doi.org/10.1038/s41598-019-47249-7). A copy of creative commons license can be found at the link http://creativecommons.org/licenses/by/4.0/

Conflict of interest

The author declares no conflict of interest.

Author details

Vishwambhar Vishnu Bhandare
Department of Biosciences and Bioengineering, Indian Institute of Technology Bombay, Mumbai, India

*Address all correspondence to: vishwayogi@gmail.com

References

[1] Bayat A. Science, medicine, and the future: Bioinformatics. BMJ [Internet]. 2002;324(7344):1018-22. Available from: http://www.ncbi.nlm.nih.gov/pubmed/11976246

[2] Melo AM, Coraor J, Alpha-Cobb G, Elbaum-Garfinkle S, Nath A, Rhoades E. A functional role for intrinsic disorder in the tau-tubulin complex. Proc Natl Acad Sci [Internet]. 2016;113(50):14336-41. Available from: http://www.pnas.org/lookup/doi/10.1073/pnas.1610137113

[3] Butner KA, Kirschner MW. Tau Protein Binds to Microtubules through. J Cell Biol. 1991;115(3):717-30.

[4] Guo T, Noble W, Hanger DP. Roles of tau protein in health and disease. Acta Neuropathol [Internet]. 2017 May;133(5):665-704. Available from: http://www.ncbi.nlm.nih.gov/pubmed/28386764

[5] Liu F, Gong C-X. Tau exon 10 alternative splicing and tauopathies. Mol Neurodegener [Internet]. 2008;3(1):8. Available from: http://molecularneurodegeneration.biomedcentral.com/articles/10.1186/1750-1326-3-8

[6] Buée L, Bussière T, Buée-Scherrer V, Delacourte A, Hof PR. Tau protein isoforms, phosphorylation and role in neurodegenerative disorders. Brain Res Rev. 2000;33(1):95-130.

[7] Kolarova M, García-Sierra F, Bartos A, Ricny J, Ripova D. Structure and Pathology of Tau Protein in Alzheimer Disease. Int J Alzheimers Dis [Internet]. 2012;2012:1-13. Available from: http://www.hindawi.com/journals/ijad/2012/731526/

[8] Kellogg EH, Hejab NMA, Poepsel S, Downing KH, DiMaio F, Nogales E. Near-atomic model of microtubule-tau interactions. Science (80-) [Internet]. 2018;1780(May):eaat1780. Available from: http://www.sciencemag.org/lookup/doi/10.1126/science.aat1780

[9] Jebarupa B, Muralidharan M, Arun A, Mandal AK, Mitra G. Conformational heterogeneity of tau: Implication on intrinsic disorder, acid stability and fibrillation in Alzheimer's disease. Biophys Chem [Internet]. 2018 Oct;241:27-37. Available from: http://www.ncbi.nlm.nih.gov/pubmed/30081240

[10] Qiang L, Sun X, Austin TO, Muralidharan H, Jean DC, Liu M, et al. Tau Does Not Stabilize Axonal Microtubules but Rather Enables Them to Have Long Labile Domains. Curr Biol [Internet]. 2018 Jul 9;28(13):2181-2189.e4. Available from: http://www.ncbi.nlm.nih.gov/pubmed/30008334

[11] Brettschneider J, Arai K, Del Tredici K, Toledo JB, Robinson JL, Lee EB, et al. TDP-43 pathology and neuronal loss in amyotrophic lateral sclerosis spinal cord. Acta Neuropathol. 2014;128(3):423-437.

[12] Gao Y-L, Wang N, Sun F-R, Cao X-P, Zhang W, Yu J-T. Tau in neurodegenerative disease. Ann Transl Med [Internet]. 2018 May;6(10):175-175. Available from: http://atm.amegroups.com/article/view/19456/19578

[13] Friedhoff P, von Bergen M, Mandelkow E-M, Mandelkow E. Structure of tau protein and assembly into paired helical filaments. Biochim Biophys Acta - Mol Basis Dis [Internet]. 2000;1502(1):122-32. Available from: http://linkinghub.elsevier.com/retrieve/pii/S0925443900000387

[14] Kellogg EH, Hejab NMA, Poepsel S, Downing KH, DiMaio F, Nogales E. Near-atomic model of microtubule-tau

interactions. Science (80-) [Internet]. 2018;eaat1780. Available from: http://www.sciencemag.org/lookup/doi/10.1126/science.aat1780

[15] Kadavath H, Cabrales Fontela Y, Jaremko M, Jaremko Ł, Overkamp K, Biernat J, et al. The Binding Mode of a Tau Peptide with Tubulin. Angew Chemie Int Ed [Internet]. 2018 Mar 12;57(12):3246-50. Available from: http://doi.wiley.com/10.1002/anie.201712089

[16] Panda D, Samuel JC, Massie M, Feinstein SC, Wilson L. Differential regulation of microtubule dynamics by three- and four-repeat tau: Implications for the onset of neurodegenerative disease. Proc Natl Acad Sci [Internet]. 2003 Aug 5;100(16):9548-53. Available from: http://www.pnas.org/cgi/doi/10.1073/pnas.1633508100

[17] Panda D, Goode BL, Feinstein SC, Wilson L. Kinetic Stabilization of Microtubule Dynamics at Steady State by Tau and Microtubule-Binding Domains of Tau. Biochemistry [Internet]. 1995 Sep 5;34(35):11117-27. Available from: http://pubs.acs.org/doi/abs/10.1021/bi00035a017

[18] Ludueña RF. A Hypothesis on the Origin and Evolution of Tubulin [Internet]. Vol. 302, International Review of Cell and Molecular Biology. Elsevier; 2013. 41-185 p. Available from: http://dx.doi.org/10.1016/B978-0-12-407699-0.00002-9

[19] Ludueña RF, Banerjee A. The Isotypes of Tubulin. In: The Role of Microtubules in Cell Biology, Neurobiology, and Oncology [Internet]. Totowa, NJ: Humana Press; 2008. p. 123-75. Available from: http://link.springer.com/10.1007/978-1-59745-336-3_6

[20] Guo J, Walss-Bass C, Ludueña RF. The beta isotypes of tubulin in neuronal differentiation. Cytoskeleton (Hoboken) [Internet]. 2010 Jul;67(7):431-41. Available from: http://www.ncbi.nlm.nih.gov/pubmed/20506160

[21] Kumbhar BV, Borogaon A, Panda D, Kunwar A. Exploring the Origin of Differential Binding Affinities of Human Tubulin Isotypes αβII, αβIII and αβIV for DAMA-Colchicine Using Homology Modelling, Molecular Docking and Molecular Dynamics Simulations. Toda T, editor. PLoS One [Internet]. 2016 May 26;11(5):e0156048. Available from: https://dx.plos.org/10.1371/journal.pone.0156048

[22] Kumbhar BV, Panda D, Kunwar A. Interaction of microtubule depolymerizing agent indanocine with different human αβ tubulin isotypes. 2018;1-20. Available from: http://journals.plos.org/plosone/article/file?id=10.1371/journal.pone.0194934&type=printable

[23] Pamula MC, Ti S-C, Kapoor TM. The structured core of human β tubulin confers isotype-specific polymerization properties. J Cell Biol [Internet]. 2016 May 23;213(4):425-33. Available from: http://www.jcb.org/lookup/doi/10.1083/jcb.201603050

[24] Fees CP, Aiken J, O'Toole ET, Giddings TH, Moore JK. The negatively charged carboxy-terminal tail of β-tubulin promotes proper chromosome segregation. Mol Biol Cell [Internet]. 2016;27(11):1786-96. Available from: http://www.molbiolcell.org/lookup/doi/10.1091/mbc.E15-05-0300

[25] Roll-Mecak A. Intrinsically disordered tubulin tails: complex tuners of microtubule functions? Semin Cell Dev Biol [Internet]. 2015 Jan;37:11-9. Available from: http://www.ncbi.nlm.nih.gov/pubmed/25307498

[26] Janke C. The tubulin code: Molecular components, readout mechanisms, functions. J Cell Biol. 2014;206(4):461-72.

[27] Pamula MC, Ti S-C, Kapoor TM. The structured core of human β tubulin confers isotype-specific polymerization properties. J Cell Biol [Internet]. 2016;213(4):425-33. Available from: http://www.ncbi.nlm.nih.gov/pubmed/27185835

[28] Panda D, Miller HP, Banerjee A, Luduena RF, Wilson L. Microtubule dynamics in vitro are regulated by the tubulin isotype composition. Proc Natl Acad Sci [Internet]. 1994 Nov 22;91(24):11358-62. Available from: http://www.ncbi.nlm.nih.gov/pubmed/7972064

[29] Shojania Feizabadi M, Janakaloti Narayanareddy BR, Vadpey O, Jun Y, Chapman D, Rosenfeld S, et al. Microtubule C-Terminal Tails Can Change Characteristics of Motor Force Production. Traffic [Internet]. 2015 Oct;16(10):1075-87. Available from: http://doi.wiley.com/10.1111/tra.12307

[30] Banerjee A, Luduena RF. Kinetics of colchicine binding to purified beta-tubulin isotypes from bovine brain. J Biol Chem. 1992;267(19):13335-9.

[31] Cowan NJ, Lewis SA, Gu W, Buraoyne RD. Tubulin Isotypes and Their Interaction with Microtubule Associated Proteins. Protoplasma. 1988;145:6-111.

[32] Murphy DB. Purification of Tubulin and Tau from Chicken Erythrocytes: Tubulin Isotypes and Mechanisms of Microtubule Assembly. Methods Enzymol. 1991;196(1986):235-46.

[33] Ludueuna RF. Are Tubulin Isotypes Functionally Significant. Mol Biol Cell. 1993;4(May):445-57.

[34] Vemu A, Atherton J, Spector JO, Moores CA, Roll-Mecak A. Tubulin isoform composition tunes microtubule dynamics. Mol Biol Cell [Internet]. 2017 Dec 1;28(25):3564-72. Available from: http://www.ncbi.nlm.nih.gov/pubmed/29021343

[35] Feizabadi MS, Hernandez MAV, Breslin JB, Akintola II. The regulatory effect of Tau protein on polymerization of MCF7 microtubules in vitro. Biochem Biophys Reports [Internet]. 2019 Mar;17:151-6. Available from: https://linkinghub.elsevier.com/retrieve/pii/S2405580818302528

[36] Bhandare VV, Kumbhar BV, Kunwar A. Differential binding affinity of tau repeat region R2 with neuronal-specific β-tubulin isotypes. Sci Rep [Internet]. 2019 Jul 25;9(1):10795. Available from: http://www.ncbi.nlm.nih.gov/pubmed/31346240

[37] Sievers F, Wilm A, Dineen D, Gibson TJ, Karplus K, Li W, et al. Fast, scalable generation of high-quality protein multiple sequence alignments using Clustal Omega. Mol Syst Biol [Internet]. 2014 Apr 16;7(1):539-539. Available from: http://msb.embopress.org/cgi/doi/10.1038/msb.2011.75

[38] Webb B, Sali A. Comparative Protein Structure Modeling Using MODELLER. In: Current Protocols in Bioinformatics [Internet]. Hoboken, NJ, USA: John Wiley & Sons, Inc.; 2016. p. 5.6.1-5.6.37. Available from: http://doi.wiley.com/10.1002/cpbi.3

[39] Biasini M, Bienert S, Waterhouse A, Arnold K, Studer G, Schmidt T, et al. SWISS-MODEL: modelling protein tertiary and quaternary structure using evolutionary information. Nucleic Acids Res [Internet]. 2014 Jul 1;42(W1):W252-8. Available from: http://academic.oup.com/nar/article/42/W1/W252/2435313/SWISSMODEL-modelling-protein-tertiary-and

[40] Eisenberg D, Lüthy R, Bowie JU. [20] VERIFY3D: Assessment of protein models with three-dimensional profiles. Methods Enzymol. 1997;277:396-404.

[41] Colovos C, Yeates TO. Verification of protein structures: patterns of nonbonded atomic interactions. Protein

Sci [Internet]. 1993 Sep;2(9):1511-9. Available from: http://www.ncbi.nlm.nih.gov/pubmed/8401235

[42] Laskowski RA, MacArthur MW, Moss DS, Thornton JM. PROCHECK: a program to check the stereochemical quality of protein structures. J Appl Crystallogr [Internet]. 1993 Apr 1;26(2):283-91. Available from: http://scripts.iucr.org/cgi-bin/paper?S0021889892009944

[43] Van Der Spoel D, Lindahl E, Hess B, Groenhof G, Mark AE, Berendsen HJC. GROMACS: fast, flexible, and free. J Comput Chem. 2005;26(16):1701-1718.

[44] Karplus M, McCammon JA. Molecular dynamics simulations of biomolecules. Nat Struct Mol Biol. 2002;9(9):646-52.

[45] Hospital A, Goñi JR, Orozco M, Gelpí JL. Molecular dynamics simulations: advances and applications. Adv Appl Bioinforma Chem AABC. 2015;10:37.

[46] Berendsen HJC, van der Spoel D, van Drunen R. GROMACS: A message-passing parallel molecular dynamics implementation. Comput Phys Commun. 1995;91(1-3):43-56.

[47] Lindorff-Larsen K, Piana S, Palmo K, Maragakis P, Klepeis JL, Dror RO, et al. Improved side-chain torsion potentials for the Amber ff99SB protein force field. Proteins Struct Funct Bioinforma [Internet]. 2010;78(8):NA-NA. Available from: http://doi.wiley.com/10.1002/prot.22711

[48] Meagher KL, Redman LT, Carlson HA. Development of polyphosphate parameters for use with the AMBER force field. J Comput Chem [Internet]. 2003 Jul 15;24(9):1016-25. Available from: http://www.ncbi.nlm.nih.gov/pubmed/12759902

[49] Allnér O, Nilsson L, Villa A. Magnesium Ion-Water Coordination and Exchange in Biomolecular Simulations. J Chem Theory Comput [Internet]. 2012 Apr 10;8(4):1493-502. Available from: http://www.ncbi.nlm.nih.gov/pubmed/26596759

[50] Case DA, Darden TA, Cheatham III TE, Simmerling CL, Wang J, Duke RE, et al. AMBER 12; University of California: San Francisco, 2012. 2012;

[51] ParmEd tool [Internet]. Available from: http://parmed.github.io/ParmEd/html/index.html

[52] Bhandare VV, Ramaswamy A. The proteinopathy of D169G and K263E mutants at the RNA Recognition Motif (RRM) domain of tar DNA-binding protein (tdp43) causing neurological disorders: A computational study. J Biomol Struct Dyn. 2017;

[53] Essmann U, Perera L, Berkowitz ML, Darden T, Lee H, Pedersen LG. A smooth particle mesh Ewald method. J Chem Phys [Internet]. 1995 Nov 15;103(19):8577-93. Available from: http://aip.scitation.org/doi/10.1063/1.470117

[54] Darden T, York D, Pedersen L. Particle mesh Ewald: An N ·log(N) method for Ewald sums in large systems. J Chem Phys [Internet]. 1993 Jun 15;98(12):10089-92. Available from: http://aip.scitation.org/doi/10.1063/1.464397

[55] Hess B, Bekker H, Berendsen HJC, Fraaije JG. LINCS: A Linear Constraint Solver for Molecular Simulations. Artic / Lett to Ed. 1977;18:1463-1472.

[56] Kabsch W, Sander C. Dictionary of protein secondary structure: pattern recognition of hydrogen-bonded and geometrical features. Biopolymers [Internet]. 1983/12/01. 1983;22(12):2577-637. Available from: http://www.ncbi.nlm.nih.gov/entrez/query.fcgi?cmd=Retrieve&db=PubMed&dopt=Citation&list_uids=6667333

[57] Humphrey W, Dalke A, Schulten K. VMD: visual molecular dynamics. J Mol Graph [Internet]. 1996/02/01. 1996;14(1):27-28,33-38. Available from: http://www.ncbi.nlm.nih.gov/entrez/query.fcgi?cmd=Retrieve&db=PubMed&dopt=Citation&list_uids=8744570

[58] Biovia Discovery studio Visualizer. Daasault Systemes BIOVIA; 2017.

[59] Pettersen EF, Goddard TD, Huang CC, Couch GS, Greenblatt DM, Meng EC, et al. UCSF Chimera--a visualization system for exploratory research and analysis. J Comput Chem. 2004;25(13):1605-1612.

[60] Hess B, van Der Spoel D, Lindahl E. Gromacs user manual version 4.5. 4. Univ Groningen, Netherl. 2010;

[61] Gilson MK, Zhou HX. Calculation of protein-ligand binding affinities. Annu Rev Biophys Biomol Struct [Internet]. 2007;36:21-42. Available from: http://www.ncbi.nlm.nih.gov/pubmed/17201676

[62] Genheden S, Ryde U. The MM/PBSA and MM/GBSA methods to estimate ligand-binding affinities. Expert Opin Drug Discov [Internet]. 2015;10(5):449-61. Available from: http://www.ncbi.nlm.nih.gov/pubmed/25835573

[63] Wong S, Amaro RE, McCammon JA. MM-PBSA Captures Key Role of Intercalating Water Molecules at a Protein–Protein Interface. J Chem Theory Comput [Internet]. 2009 Feb 10;5(2):422-9. Available from: http://pubs.acs.org/doi/abs/10.1021/ct8003707

[64] Aldeghi M, Bodkin MJ, Knapp S, Biggin PC. Statistical Analysis on the Performance of Molecular Mechanics Poisson–Boltzmann Surface Area versus Absolute Binding Free Energy Calculations: Bromodomains as a Case Study. J Chem Inf Model [Internet]. 2017 Sep 25;57(9):2203-21. Available from: http://pubs.acs.org/doi/10.1021/acs.jcim.7b00347

[65] Kant V, Vijayakumar S, Sahoo GC, Chaudhery SS, Das P. In-silico screening and validation of high-affinity tetra-peptide inhibitor of Leishmania donovani O-acetyl serine sulfhydrylase (OASS). J Biomol Struct Dyn [Internet]. 2018 Feb 7;1-14. Available from: https://www.tandfonline.com/doi/full/10.1080/07391102.2018.1429315

[66] Paissoni C, Spiliotopoulos D, Musco G, Spitaleri A. GMXPBSA 2.0: A GROMACS tool to perform MM/PBSA and computational alanine scanning. Comput Phys Commun. 2014;185(11):2920-9.

[67] Bhandare VV, Ramaswamy A. Structural dynamics of human argonaute2 and its interaction with siRNAs designed to target mutant tdp43. Adv Bioinformatics. 2016;2016.

[68] Kumari R, Kumar R, Lynn A. g_mmpbsa-A GROMACS Tool for High-Throughput MM-PBSA Calculations. J Chem Inf Model. 2014;54(7):1951-1962.

[69] Cavuturu BM, Bhandare VV, Ramaswamy A, Arumugam N. Molecular dynamics of interaction of Sesamin and related compounds with the cancer marker β-catenin: an in silico study. J Biomol Struct Dyn. 2018;

[70] Guo W, Chen Y, Zhou X, Kar A, Ray P, Chen X, et al. An ALS-associated mutation affecting TDP-43 enhances protein aggregation, fibril formation and neurotoxicity. Nat Struct Mol Biol [Internet]. 2011/06/15. 18(7):822-30. Available from: http://www.ncbi.nlm.nih.gov/entrez/query.fcgi?cmd=Retrieve&db=PubMed&dopt=Citation&list_uids=21666678

[71] Wang C, Greene D, Xiao L, Qi R, Luo R. Recent Developments and

Applications of the MMPBSA Method. Front Mol Biosci [Internet]. 2017;4:87. Available from: http://www.ncbi.nlm.nih.gov/pubmed/29367919

[72] Miller BR, McGee TD, Swails JM, Homeyer N, Gohlke H, Roitberg AE. MMPBSA.py: An Efficient Program for End-State Free Energy Calculations. J Chem Theory Comput [Internet]. 2012 Sep 11;8(9):3314-21. Available from: http://www.ncbi.nlm.nih.gov/pubmed/26605738

[73] Ti S-C, Alushin GM, Kapoor TM. Human β-Tubulin Isotypes Can Regulate Microtubule Protofilament Number and Stability. Dev Cell [Internet]. 2018 Sep; Available from: https://linkinghub.elsevier.com/retrieve/pii/S1534580718306841

[74] Biasini M, Bienert S, Waterhouse A, Arnold K, Studer G, Schmidt T, et al. SWISS-MODEL: Modelling protein tertiary and quaternary structure using evolutionary information. Nucleic Acids Res. 2014;42(W1).

[75] Wiederstein M, Sippl MJ. ProSA-web: interactive web service for the recognition of errors in three-dimensional structures of proteins. Nucleic Acids Res [Internet]. 2007 Jul;35(Web Server issue):W407-10. Available from: http://www.ncbi.nlm.nih.gov/pubmed/17517781

[76] Lovell SC, Davis IW, Arendall WB, de Bakker PIW, Word JM, Prisant MG, et al. Structure validation by Calpha geometry: phi,psi and Cbeta deviation. Proteins [Internet]. 2003 Feb 15;50(3):437-50. Available from: http://www.ncbi.nlm.nih.gov/pubmed/12557186

[77] RAMACHANDRAN GN, RAMAKRISHNAN C, SASISEKHARAN V. Stereochemistry of polypeptide chain configurations. J Mol Biol [Internet]. 1963 Jul;7:95-9. Available from: http://www.ncbi.nlm.nih.gov/pubmed/13990617

[78] Van Der Spoel D, Lindahl E, Hess B, Groenhof G, Mark AE, Berendsen HJ. GROMACS: fast, flexible, and free. J Comput Chem [Internet]. 2005/10/08. 2005;26(16):1701-18. Available from: http://www.ncbi.nlm.nih.gov/entrez/query.fcgi?cmd=Retrieve&db=PubMed&dopt=Citation&list_uids=16211538

[79] Kar S, Fan J, Smith MJ, Goedert M, Amos LA. Repeat motifs of tau bind to the insides of microtubules in the absence of taxol. EMBO J [Internet]. 2003 Jan 2;22(1):70-7. Available from: http://www.ncbi.nlm.nih.gov/pubmed/12505985

[80] Chau M-F, Radeke MJ, de Inés C, Barasoain I, Kohlstaedt LA, Feinstein SC. The Microtubule-Associated Protein Tau Cross-Links to Two Distinct Sites on Each α and β Tubulin Monomer via Separate Domains †. Biochemistry [Internet]. 1998 Dec;37(51):17692-703. Available from: http://pubs.acs.org/doi/abs/10.1021/bi9812118

[81] Al-Bassam J, Ozer RS, Safer D, Halpain S, Milligan RA. MAP2 and tau bind longitudinally along the outer ridges of microtubule protofilaments. J Cell Biol. 2002;157(7):1187-96.

[82] Santarella RA, Skiniotis G, Goldie KN, Tittmann P, Gross H, Mandelkow E-M, et al. Surface-decoration of microtubules by human tau. J Mol Biol [Internet]. 2004 Jun 4;339(3):539-53. Available from: http://www.ncbi.nlm.nih.gov/pubmed/15147841

[83] Luo Y, Ma B, Nussinov R, Wei G. Structural Insight into Tau Protein's Paradox of Intrinsically Disordered Behavior, Self-Acetylation Activity, and Aggregation. J Phys Chem Lett [Internet]. 2014 Sep 4;5(17):3026-31. Available from: http://www.ncbi.nlm.nih.gov/pubmed/25206938

[84] Avila J, Jiménez JS, Sayas CL, Bolós M, Zabala JC, Rivas G, et al. Tau Structures. Front Aging Neurosci [Internet]. 2016;8:262. Available from: http://www.ncbi.nlm.nih.gov/pubmed/27877124

[85] Fischer D, Mukrasch MD, Von Bergen M, Klos-Witkowska A, Biemat J, Griesinger C, et al. Structural and microtubule binding properties of tau mutants of frontotemporal dementias. Biochemistry. 2007;46(10):2574-82.

[86] Ma B, Wei G, Zhen J, Nussinov R. Dancing with Strings: The Conformational Dynamics of VQIXXK Motifs within Tau Protein in Monomer, Fibril and Hyper-Phosphorylated Filament States. Biophys J [Internet]. 2016;110(3):553a-554a. Available from: http://linkinghub.elsevier.com/retrieve/pii/S0006349515041430

Section 4

Applications

Chapter 6

Design of Bioelectrochemical Interfaces Assisted by Molecular Dynamics Simulations

Abraham Vidal-Limon, Guillermo Antonio Huerta-Miranda, Wendy I. García-García and Margarita Miranda-Hernández

Abstract

The design of bioelectrochemical interfaces (BEI) is an interesting topic that recently demands attention. The synergy between biomolecules and chemical components is necessary to achieve high molecular selectivity and sensitivity for the development of biosensors, synthesis of different compounds, or catalytic processes. For most BEI, the charge transfer process occurs in environments with particular chemical conditions; modeling these environments is a challenging task and requires multidisciplinary efforts. These interfaces can be composed of biomolecules, such as proteins, DNA, or more complex systems like microorganisms. Oxidoreductases enzymes are good candidates, among others, due to their catalytic activities and structural characteristics. In BEI, enzymes are immobilized on conductive surfaces to improve charge transfer processes. Covalent immobilization is the most common method to prolong lifetime or modulate the detection process. However, it is necessary to implement new methodologies that allow the selection of the best candidates for a more efficient design. Homology modeling of oxidoreductases combined with Molecular Dynamics (MD) simulation methods are alternative and already routinely used tools to investigate the structure, dynamics, and thermodynamics of biological molecules. Our motivation is to show different techniques of molecular modeling (Homology Modeling, Gaussian accelerated molecular dynamics, directed adaptive molecular dynamics and electrostatic surface calculations), and using horseradish peroxidase as a model to understand the interactions between biomolecules and gold nanoclusters (as current collector). Additionally, we present our previous studies considering molecular simulations and we discuss recent advances in biomolecular simulations aimed at biosensor design.

Keywords: bioelectrochemical interfaces, homology modeling, covalent immobilization, gold nanoclusters, molecular dynamics

1. Introduction

For some years now, the design and construction of bioelectrochemical interfaces (BEI), ranging from electrochemical biosensors (EC) for analytical applications, biofuel cells (BFC) to the development of biocomputing systems for information processing, have been topics where the scientific community has considerable participation. The implementation and integration of biomolecules with electronic elements or conductive surfaces to produce bioelectronic devices is a current topic and gathers great importance for developing biosensors that have an essential role in clinical applications, food quality control analysis and forensic medicine [1–3]. These applications always involve high sensitivity, low-sample volume, and low-cost production. In most BEI, charge transfer processes are carried out in environments with high ionic concentration, necessary for their function. These interfaces can be composed of macromolecules such as proteins, peptides, or more complex systems such as bacteria. Since the activity and lifetime depend on the correct interaction between the conductive surfaces (current collector) and the biomaterials, to achieve good bioelectrochemical responses, the enzymes need to improve the orientation towards the surface of the current collector [4, 5].

1.1 Covalent immobilization through alkanethiol linkers

The immobilization of various biomaterials is an important issue for BEI design, the covalent immobilization of enzymes stands out among the different enzymatic coupling strategies since the bonds formed through the linkers anchored to the current collector promotes direct charge transfer responses [6, 7]. Alcakanothiols are organic molecules widely used to establish self-assembled monolayers (SAMs) on the surface of gold electrodes [8–11]. The thiol groups chemisorb on gold electrodes forming strong thiolate-gold bonds (Au-S). The resulting monolayer may be used to design molecular scaffolds to couple enzymes [8, 12]. The main drawback of covalent immobilization systems is the risk of having a low surface concentration of enzymes that are either active or correctly oriented for direct charge transfer; which may lead to a low-efficiency bioelectrochemical response [13]. Therefore, having active enzymes and a good bioelectrochemical response becomes an important task. Aromatic molecules like 4-aminothiophenol, structurally mimics enzymatic substrates and can promote stable and direct contacts, necessary for efficient bioelectrochemical reactions.

Researches have proposed nanomaterials as support matrices for the design of BEI to increase the surface concentration of active enzymes on the current collector. During the last decade, nanomaterials, coupled with enzymes, have had significant relevance in the design of biosensors [5, 14–16]. Current advances in synthesis methodologies of these materials allow having a wide variety of nanomaterials with different sizes, shapes, surface charges, and physicochemical characteristics [4, 14, 17].

Some of these nanomaterials can be modified in different ways to improve biocompatibility. Biocompatibility refers to the biological nature events that do not interfere with those of electronic signal transduction and vice versa. Metallic nanoparticles are promising materials that increase the electroactive area and improve the sensitivity and stability of the attached enzymes on the electrodes, bringing the enzymatic active site close to the electrode (the redox cofactor should not exceed a 20 Å distance) to achieve direct charge transfer reactions [14, 18–23].

In our group, we have established the conditions to improve the electrochemical response of horseradish peroxidase (HRP) coupled to gold nanoparticles modified with the aromatic alkanothiol 4-aminotiophenol, using glassy carbon (GC) as current collector [24]. We proposed that a critical factor in HRP bioelectrochemical response is to promote conformational changes and proper enzyme orientation when coupled to the electrode. The protein conformation around the prosthetic group will determine the redox potential of the Fe^{3+}/Fe^{2+} couple. The activity towards H_2O_2 also depends on the protein structural arrangements [25, 26]. We found that a proper environment for the enzyme activity was achieved by increasing the distance between gold nanoparticles. The voltammetric studies of the prosthetic heme group showed significant differences between the enzyme immobilized on randomly deposited gold nanoparticles and the enzyme immobilized on a well-dispersed gold nanoparticles deposit (**Figure 1**). The dispersed electrode improved the electrochemical response of the enzyme; this fact showed that the distance between enzymes is important, probably because longer distances decreased the steric impediments between the enzymes, and it appears to be a better immobilization strategy. With these results, we proposed that an essential part of BEI design is related to the structure of the biological systems. The electrochemical sensing is dependent on the recognition or transformation of the substrate, as previously observed [27].

The advent of high-resolution and robust techniques as X-ray crystallography [28, 29] and cryo-EM [30–32] has enormously contributed to the baggage of structural information available for bioelectrochemical applications. These techniques describe the atomic positions of enzymes, as well as the conformational dynamics resulted from recognition and binding processes, allowing the prediction of the possible electrochemical behavior of enzymes immobilized on electrodes.

1.2 Molecular modeling methodologies

The use of computational tools such as molecular mechanics and quantum mechanics for the study of chemical and biological reactions, involves a mathematical treatment of a large number of particles (hundreds of thousands of atoms that build up molecules like proteins, nucleic acids, lipids, and sugars) [33, 34]. In general terms, these methods are ideal for obtaining chemical and physical properties from three-dimensional molecular models [35–37]. It is necessary to correlate the structural models with their corresponding chemical, catalytic or biological properties; towards the rationalization and design of molecules for specific applications. Mainly, molecular Dynamics (MD) methods are based on the solution of Newton's second law for all the atoms in the system, and help to predict the behavior of a biomolecule during a specific time. The integration of motion equations allows to analyze the trajectories corresponding to position, velocity, and acceleration of each particle of the system in any

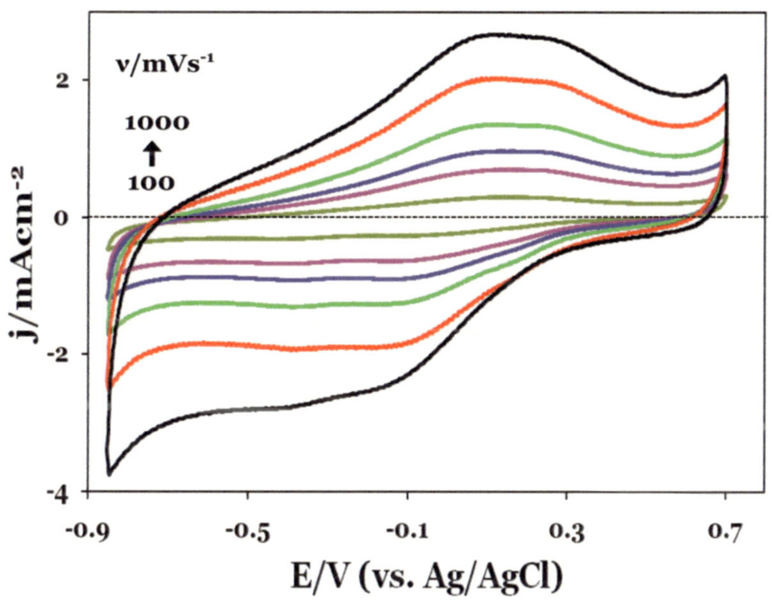

Figure 1.
Cyclic voltammogram of HRP response at different scan rates, immobilized on GC electrode modified with gold nanoparticles functionalized with the alkanethiol molecule 4-aminothiophenol. Phosphate buffer pH 6.8.

fraction of time [38]. Unfortunately, the simplification of considering atoms as spheres, and bonds as springs or tensors (with different parameters depending on the type of bond), makes difficult to study bond-breaking and formation, or reaction mechanisms. Quantum mechanics (QM) mathematically describes the fundamental behavior of matter on an electronic scale by solving the Schrödinger equation for each atom in the system. QM also describes the behavior of atoms and molecules, in terms of their chemical reactivity, geometry, and their optical, electrical, magnetic, and mechanical properties [39, 40]. The coupled implementation of these techniques could describe the biomolecules involved in charge transfer processes, and design strategies for efficient BEI.

1.3 Molecular dynamics as a tool to predict the electrochemical activity in proteins

A challenging task in BEI design is the selection of robust biomolecules capable of tolerate non-physiological conditions, without loosing efficiency and conformation [2]. The protein folding is governed by a series of molecular interactions between the amino acids and the surrounding chemical environment [34]. Therefore, predicting their behavior with molecular models contributes to the optimization of resources before the development of BEI.

In our previous work the VP6 rotavirus capsid protein was encapsulated with the ionic polymer Nafion on GC. We demonstrated that this electrode could transfer electric charge applying an external potential, when using the redox probe potassium ferricyanide ($K_3[Fe(CN)_6]$). In parallel, we modeled by MD the electrostatic and conformational states of VP6 (**Figure 2**). This model suggested a route for ionic conductivity, where the electrostatic protein surface displayed a negative charge which interacted with the ferricyanide redox probe, promoting the charge transfer reaction. In order to show if this electrochemical activity was particular to proteins in general, under the same conditions as the experiments with VP6, bovine serum albumin (BSA), whose primary biological function is

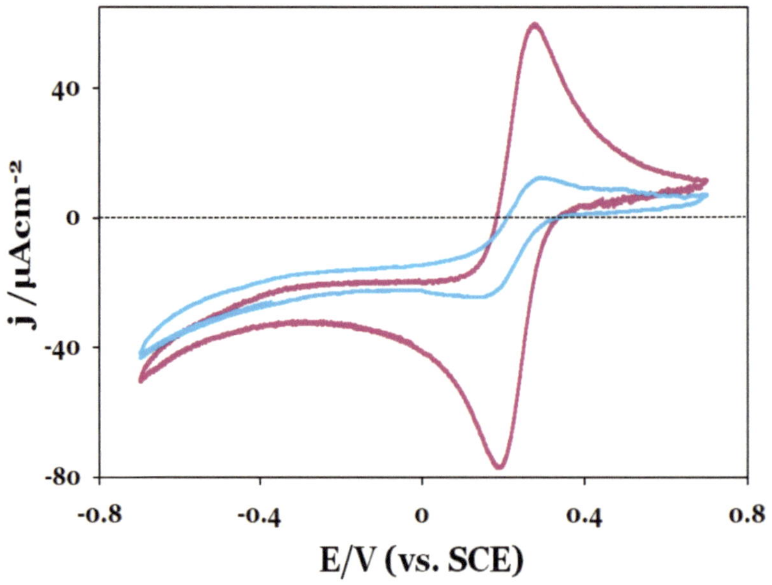

Figure 2.
Comparison of the cyclic Voltammogram of VP6 (pink) and BSA (blue) coupled to GC, at 20 mV/s in 0.01 M $K_3[Fe(CN)_6]$, 1 M KCl.

to transport different molecules through the bloodstream, was tested. However, we did not observe the charge transfer reaction of potassium ferricyanide when this protein was encapsulated (**Figure 2**). These results demonstrated that VP6 protein could be used as conductive scaffold for the development of different BEI applications [41].

Calculations of polarized electrostatic surface potentials from homology models of viral proteins, surprisingly showed that several capsids could display field effects as the recorded on VP6. In contrast, this effect cannot be displayed on non-charged proteins as BSA (**Figure 3**). Therefore, our previous data on viral capsids can serve as a workflow to identify candidate proteins or enzymes for construction of BEI devices. Altogether, theoretical and experimental reports on BEI [41–43] are examples of how structural information could help to elucidate the behavior of biological systems during electrochemical reactions. Nonetheless, the aforementioned studies depend on structural information from biomolecules, and certain biomolecules are very complex and highly labile, which difficult elucidation of atomic positions. However, since new protein sequences are continuously available, predictive methodologies as homology modeling becomes a valuable tool to asset structural information.

The aim of this multidisciplinary research is to perform molecular simulations on horseradish peroxidase as a model system, to generate data that can be used as selection criteria for design and further experimental validation in BEI development.

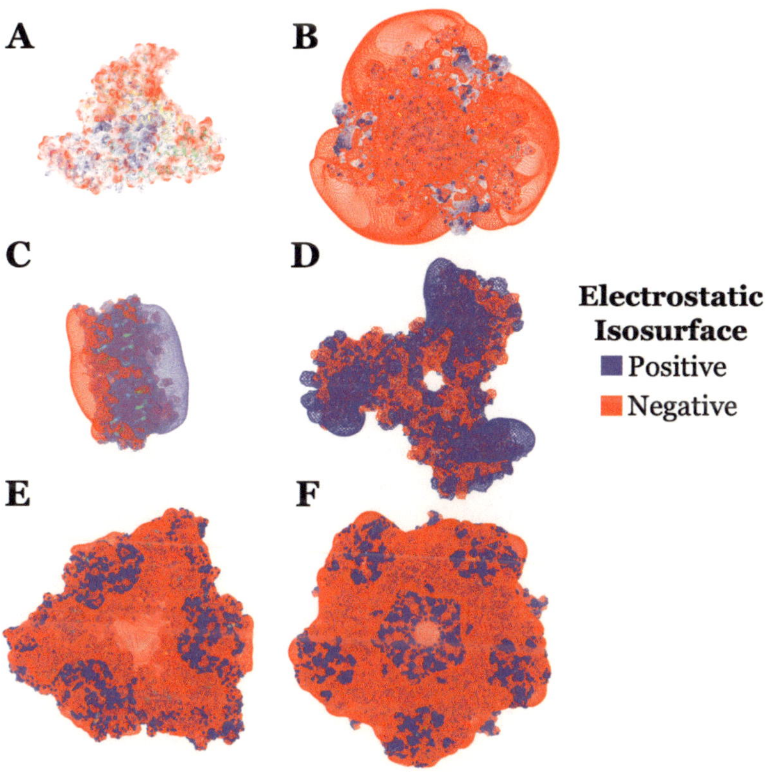

Figure 3.
Polarized electrostatic Isosurface potentials of different viral capsids. (A) Bovine serum albumin protein, (B) rotavirus VP6 trimer capsid protein, (C) influenza virus capsid protein, (D) hepatitis B virus capsid protein, (E) HIV-1 virus capsid protein, (F) HPV virus capsid protein. All polarized electrostatic Isosurfaces were calculated with PME approximation on capsid arrangements and displayed as positive (blue) or negative (red) mesh isosurfaces with VMD v.1.9.3 [44].

2. General techniques: Horseradish modeling and system building

The X-ray crystal structure from horseradish peroxidase (HRP) was collected from Protein Data Bank (www.rcsb.org/, PDB code 1H5A) [45], and reconstructed using i-tasser metaserver [46]. The best model was selected based on lowest RMSD and higher C-value, to achieve the best initial coordinates. To improve the model system, all missing atoms and sidechain positions were modeled against available plant peroxidases on the PDB. A penta-coordinated resting state heme group was crystallized as Fe^{3+}, and axial coordination with a histidine groups was retained.

The Protein Preparation Wizard routine (Schrödinger Maestro v2019-4, New York, 2019), was applied as a preparation step to correctly assign missing hydrogen atoms and protonation states of ionizable residues. Finally, PROPKA2.0 sub-routine was applied to whole model system [47] and pH 7 was fixed accordingly to our previous studies.

Enzyme structural parameters were described using the ff14SB force field [48]. The full systems comprised ~60,000 atoms in a cubic box of 15 Å length including TIP3P explicit water and Na + ions to ensure overall charge neutrality. Non-bonded interactions were calculated within a 12 Å cutoff, and long-range electrostatics were treated using the Particle-Mesh Ewald method [49]. SHAKE algorithm was enabled to constrain all bonds involving hydrogen during simulations. Conventional MD protocol for each system replica comprised: 1) 5000 steepest descent minimization steps followed by 10,000 conjugate gradient minimization steps; 2) 500 ps of progressive NVT heating from 0 to 300 K; 3) 5 ns of NVT equilibration, and 4) 10 ns of NPT dynamics at 300 K and 1 bar.

2.1 Gaussian accelerated molecular dynamics (GaMD)

Conventional MD simulations were carried out to initially relax full system coordinates, using AMBER18 software package [50]. Then, the GaMD module implemented in AMBER v. 18 was applied to perform extra 50 ns of short classic MD simulation to collect the statistics for calculating GaMD acceleration parameters. Extra 50 ns of short equilibration was applied after adding the boost potential, and finally three independent 500 ns GaMD production simulations with randomized initial atomic velocities. This method applies Gaussian functions (boost potentials) on the total potential energy term, which disturb the potential energy surface allowing the exploration of different energy states. In addition, functions that directly disturb the dihedral angles of the amide bonds were applied, promoting conformational changes. All GaMD simulations were performed with a dual-boost level by setting the reference energy to the lower bound. One boost potential is applied to the dihedral energetic term and the other to the potential energy term. The average and SD of the system potential energies were calculated every 500,000 steps (1 ns) for all simulation systems. The upper limit of the boost potential SD, $\sigma 0$, was set to 6.0 $kcal{\cdot}mol^{-1}$ for both the dihedral and the total potential energetic terms [51–53]. The system temperature was ~298 K and 1 atm pressure, with integration steps of 2 fs. Three different system coordinates were extracted from GaMD trajectories, each one representing a different energy and conformational state.

2.2 Building of nanomaterial structures

The selected nanomaterials were Au nanoclusters (AuNCs) modified with two types of linkers: 4-aminothiophenol (ATP) and mercaptobenzoic acid (MBA); each cluster had 2.1 nm diameter and a nucleus composed of 96 Au atoms (Au314 (SR) 96). The nanostructures were modeled as reported elsewhere [54]. Six gold nanoclusters (6 AuNCs) were placed as a thin layer, emulating our previous experimental system (~ 20 nm gold nanoparticles) [24].

2.3 Molecular coupling

Adaptive Steered Molecular Dynamics (ASMD) simulations [55, 56] started from the earlier selected conformational states, derived from the GaMD analysis. Due to the chemical complexity of the enzyme surface, the simulations were driven by the electrostatic surface (PME) [49] for each conformational state and the 6 AuNCs layer. Hence, the scans were performed at speeds of 10 Å / ns, which comprised a 10-step profile of 1 Å between the Fe(III) of the active site and the ATP amino group or MBA group carboxylic of the 6 AuNCs. Three independent replicas were performed through gradual scans, involving measurements of the free energy of coupling between HRP and 6 AuNCs. The free energy was measured by shortening the distance between the central Fe(III) atom and the amino groups of the linkers, i.e., ATP amino groups and central Fe(III) atom of HRP heme ($NH_3^+ \rightarrow Fe^{3+}$).

3. Results and discussion

3.1 HRP model system

The direct applications for protein structures produced by molecular modeling techniques such as homology modeling, include identification of structural and functional regions within a protein, and can be exploted as a lead for further experimental studies such as mutation analysis, catalytic and electrochemical characterization [27]. Furthermore, if homology modeling is combined with other computational methods such as molecular dynamics or quantum mechanics, the produced model can be used to screen different applications.

In our case, the available information comes from the genetic sequence of HRP enzyme and incomplete structures deposited on PDB, hence, it is necessary to reconstruct and evaluate our final model system based on homology models.

Our initial model was reconstructed with i-tasser metaserver, using as template the resting state horseradish peroxidase (1H5A), and building a waterbox for further molecular dynamics analysis (**Figure 4**).

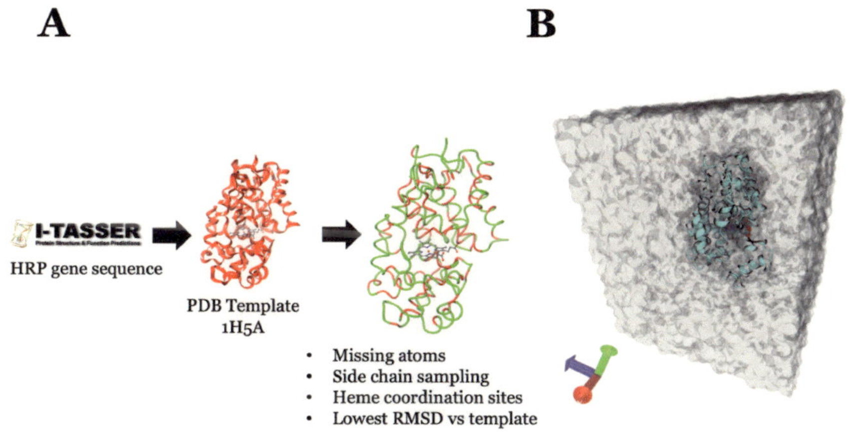

Figure 4.
Horseradish peroxidase homology model reconstruction and MD system. (A) Modeling pipeline applied to reconstruction of HRP, (B) final solvated HRP model for MD simulations.

3.2 Conformational sampling

The homology model system of HRP was subjected to an initial conformational search with GaMD method for 500 ns of simulation which is comparable to conventional MD simulations on the order of microseconds. This data allow us to reconstruct an energy hypersurface from three structural and energetic variables: total energy of protein, RMSD of backbone and superficial hydrogen bonds count. The total energy of the proteins indicates how close is our model system to a minimum energy state; RMSD of the alpha carbons is related to changes in relative positions of the atoms, regarding to the X-ray crystal model (**Figure 5**); and the hydrogen bonds count between aspartic acid (D), glutamic acid (E), lysine (K) and arginine (R) residues, indicate how prone the enzyme is to bond to AuNCs linkers functional groups (NH_3^+ or COOH). With this approach, energetically more favorable conformations were selected under two criteria: (1) Low RMSD values, *i.e.*, preserved HRP structure necessary for catalysis; (2) High solvent exposure of groups necessary for the esterification reaction, i.e., selecting those conformations where the probability of having the previous residues exposed to the solvent is higher. At the end of this sampling, three HRP conformations with the necessary intermolecular interactions were selected; with this approach we try to theoretically predict the formation of the amide bond in the solvent exposed residues, while the over-all enzyme structure remains correctly folded (**Figure 5**).

Using the GaMD approach, we were able to explore the conformational diversity along different reaction coordinates of the HRP enzyme, and to predict more efficient couplings to AuNCs. This approximation over the boost potentials allows a reconstruction with less noise on the potential energy, which results in a more robust method to deal with artifacts during variations in the potential energy [57, 58]. Hence, our data suggest that during simulation, there are different enzyme conformations which preferentially bind to AuNCs and increase the interactions with the linker functional groups.

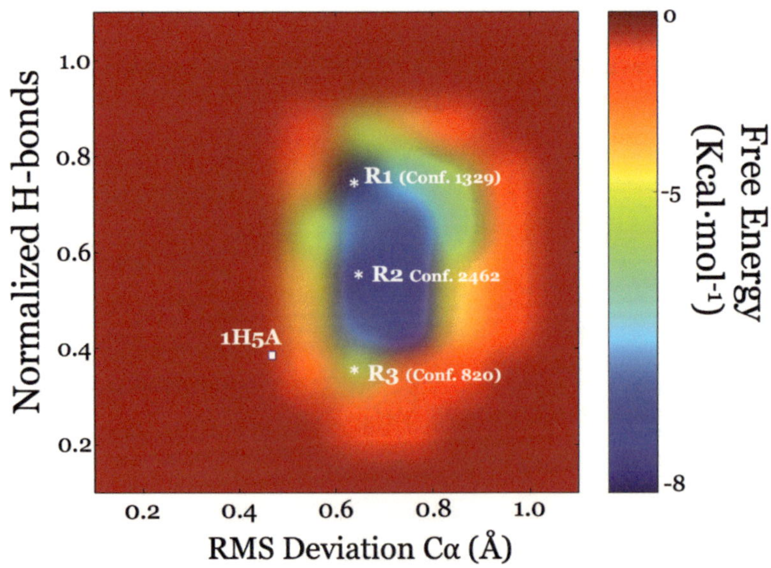

Figure 5.
Conformational sampling of HRP. The hypersurface was reconstructed using the second order McLauren cumulative expansion method. The free energy values were extracted from each conformation, depending on its structural variation and the solvent-accessible hydrogen bonds of amino acids D, E, K, and R.

3.3 Interaction between modified 6 AuNCs and HRP

The main objective of this work is to give insights using molecular modeling methodologies, for the design of more efficient BEI, based on HRP enzyme and AuNCs. The structural description of HRP with molecular dynamics methodologies can help to design experimental strategies to increase the success of the immobilization and simultaneously preserve structure and reactivity of the enzyme.

In order to estimate the immobilization efficiency between GaMD selected structures of HRP and AuCNs functionalized with linkers, free energy calculations were performed with adaptive steered molecular dynamics. The prediction model system is intended to recreate the interface previously described in Section 1.2 (**Figure 1**). GC was used as current collector and HRP was coupled to functionalized-gold nano particles modified with 4-aminothiophenol (**Figure 6A**) [24].

For all the calculations, the average electrostatic surface potential was directly extracted from GaMD, since electrostatic term is computed for each integration step. HRP showed two well characterized electrostatic isosurfaces, with a prominent

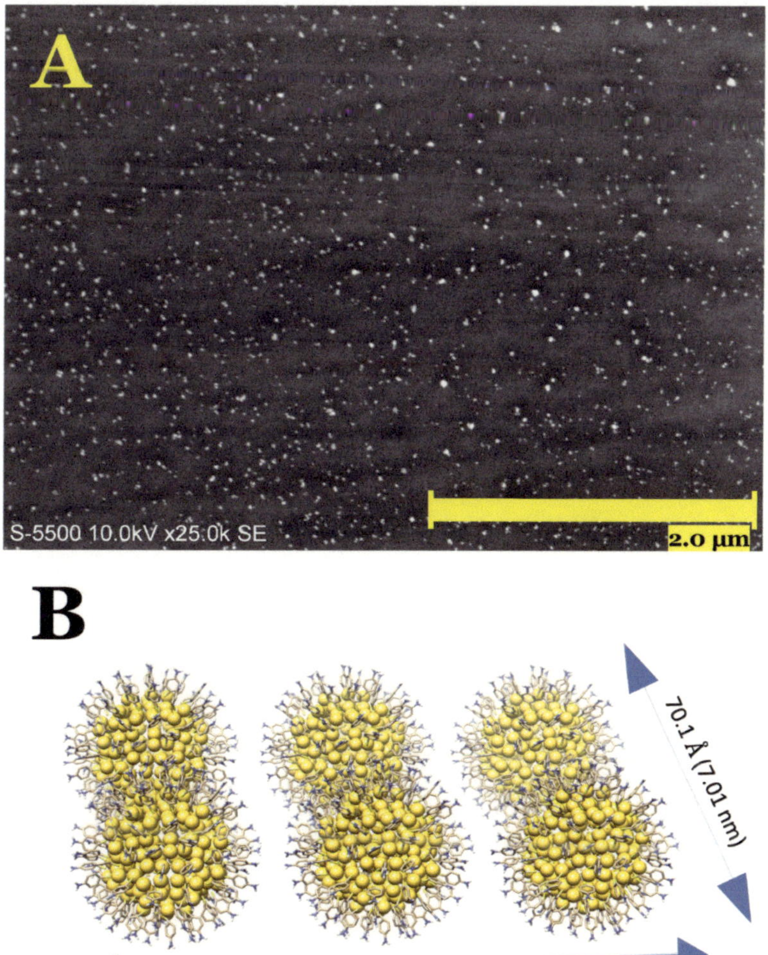

Figure 6.
Gold nanoparticles experimental and theoretical model systems. (A) SEM image of gold nanoparticle electrodeposit (average particle diameter ≈ 18 nm) in glassy carbon electrode (GC), (B) AuNCs modified with 4-aminothiophenol linker used for free energy calculations.

negatively charged surface derived from the presence of 12 exposed acidic residues. Each electrostatic potential was plotted as isosurface over each HRP conformation.

The free energy calculations between negatively charged isosurface of HRP and AuNCs modified with ATP linker (positively charged NH_3^+) showed coupling values of 60 Kcal · mol^{-1}, with a minimum distance of 22 Å from the Fe(III) of the heme group (~ 2 Å to the AuNCs NH_3^+ functional group) (**Figure 7A**). On the other hand, the positively charged isosurface of HRP showed higher coupling energies of ~120 Kcal · mol^{-1} at a minimum distance of ~21.5 Å to the same AuNCs functional group. For the AuNCs modified with MBA, the coupling energy of the negatively charged isosurface of HRP was 130 Kcal · mol^{-1} at 33 Å distance, (~3 Å from surface functional groups), while the positive HRP surface showed free coupling energies of ~75 Kcal · mol^{-1} and minimum distances of 32.5 Å from Fe(III) heme group (3.5 Å from surface functional group) (**Figure 7B**).

The difference between both HRP isosurfaces with the positively charged surface imposed by ATP was ~70 Kcal · mol^{-1} which means that the enzyme electrostatic potential imposed a clear effect for the coupling to AuNCs. However, an overall difference of 15 Kcal · mol^{-1} between both linkers showed that the coupling assays were energetically more stable for the negative isosurfaces of the HRP enzyme than the positively charged.

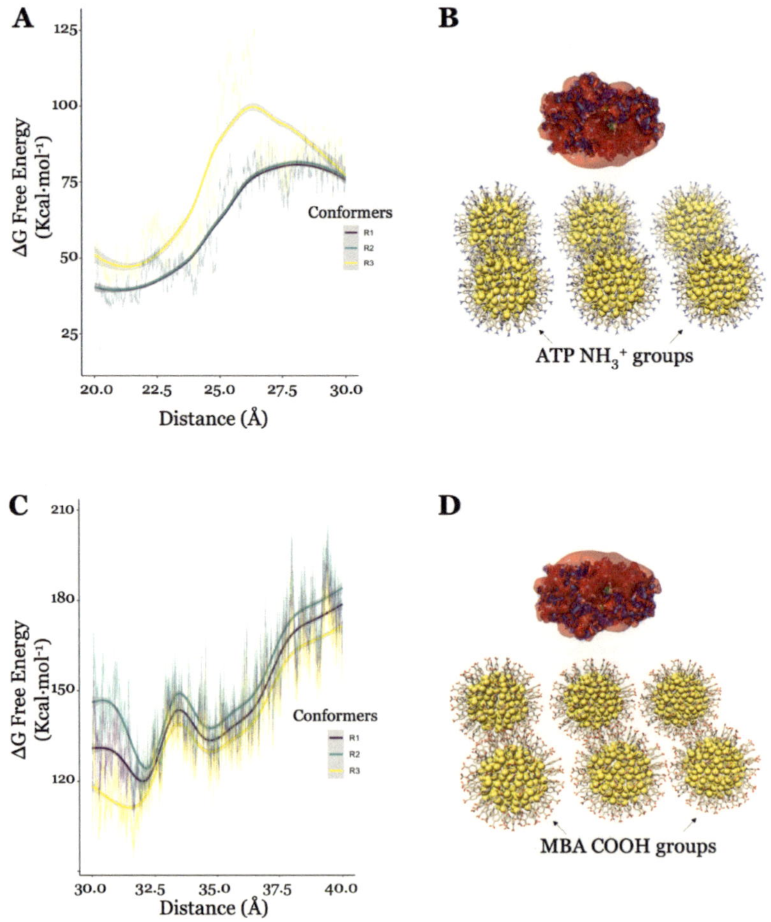

Figure 7.
Coupling free energy profiles of HRP isosurfaces and AuNCs. (A) Free energy profile of HRP coupled to 6 AuNCs-ATP, (B) molecular interaction model of negative isosurface of HRP and 6 AuNCs-ATP, (C) free energy profile of HRP coupled to 6 AuNCs-MBA, (D) positive isosurface of HRP and 6 AuNCs-MBA.

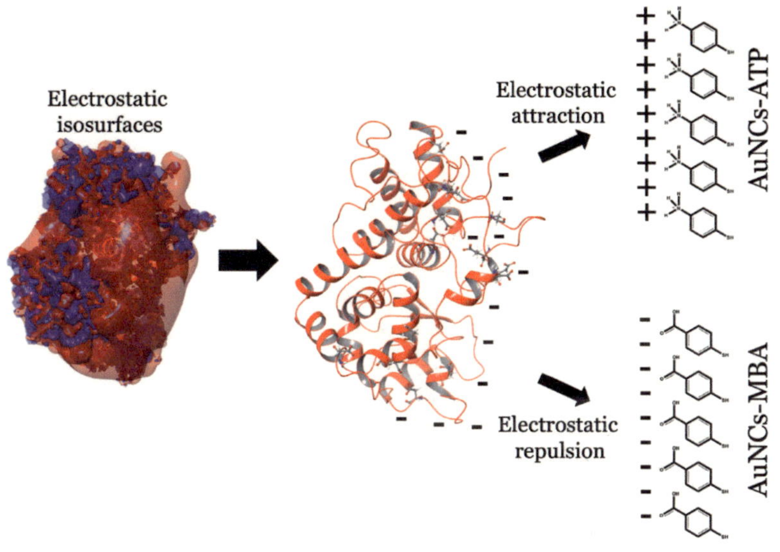

Figure 8.
Different interactions displayed between HRP and AuNCs-linkers scaffolds.

The free energy calculations suggested that HRP showed an energy minimum around the bond distance between the carboxyl groups and the amino groups [59, 60]. Our data suggest that the intermolecular interactions guided by negative electrostatic surfaces of HRP, are of lower free energy for the positive charged AuNCs-ATP, than the negative charged AuNCs-MBA (**Figure 8**). The high electrostatic attraction between HRP and AuNCs-ATP would promote an efficient electrochemical response. On the other hand, the aforementioned interactions between HRP and negative charged linkers, like MBA, are less stable and result in higher free energy coupling profiles and lower electrostatic attraction for this interfaces.

4. Conclusions

The findings of this study on HRP coupled to gold nanoclusters, indicate that the polarized electrostatic isosurface potentials are key factors to select the most efficient linker for coupling. The resulted interaction energy and distance between HRP and AuNCs-ATP are adequate to promote the formation of covalent bonds between acidic residues and amino functional groups. The evidence from this study points towards the idea that molecular simulation methods, such as homology modeling and molecular dynamics are valuable tools to take into account for design of BEI. Our previous multidisciplinary work of VP6 capsids has led us to conclude that molecular dynamics simulations elucidate structural determinants to understand the behavior of biomolecules on BEI. An implication for using homology models coupled to molecular dynamics, is the possibility of widely sample the conformational space of biomolecules probes before electrochemical experimentation. Further theoretical and experimental studies are necessary to describe the interaction with other functional group linkers and validate by electrochemical techniques the real effect of the charge difference between AuNCs-ATP and AuNCs-MBA on the redox response of this enzyme, respectively.

Acknowledgements

The authors acknowledge the financial support of CONACYT A1-S-15877, PAPIIT-DGAPA-UNAM-IN204218 and Centro Nacional de Supercomputo-IPICyT,

A.C. (National Supercomputing Center-CNS) grants for facilities at Thubat-Kaal 2 machine; very special thanks to Engineer Jesus Alaniz (CNS-IPICyT A.C) and Engineer Rogelio Elvira Morán (IER-UNAM) for their technical assistance (supercomputing and SEM analysis, respectively). W.I. García-García and G.A. Huerta-Miranda thank to CONCAYT for the PhD scholarship. A. Vidal-Limon thanks to CONACYT A1-S-15877 for the financial support of his academic fellowship.

Author details

Abraham Vidal-Limon, Guillermo Antonio Huerta-Miranda, Wendy I. García-García and Margarita Miranda-Hernández*
Institute of Renewable Energies, National Autonomous University of Mexico, Temixco, Morelos, Mexico

*Address all correspondence to: mmh@ier.unam.mx

References

[1] Calabrese Barton S, Gallaway J, Atanassov P. Enzymatic Biofuel Cells for Implantable and Microscale Devices. Chem Rev. 2004;104(10):4867-86. DOI:10.1021/cr020719k

[2] Cracknell JA, Vincent KA, Armstrong FA. Enzymes as Working or Inspirational Electrocatalysts for Fuel Cells and Electrolysis. Chem Rev. 2008;108(7):2439-61. DOI:10.1021/cr0680639

[3] Szaciłowski K. Digital Information Processing in Molecular Systems. Chem Rev 2008;108(9):3481-548. DOI:10.1021/cr068403q

[4] Jia X, Dong S, Wang E. Engineering the bioelectrochemical interface using functional nanomaterials and microchip technique toward sensitive and portable electrochemical biosensors. Biosens & Bioelectron. 2016;76:80-90. DOI: 10.1016/j.bios.2015.05.037

[5] Zhou M, Dong S. Bioelectrochemical Interface Engineering: Toward the Fabrication of Electrochemical Biosensors, Biofuel Cells, and Self-Powered Logic Biosensors. Acc Chem Res . 2011;44(11):1232-43. DOI:10.1021/ar200096g

[6] Nunes G., Marty JL. (2006) Immobilization of Enzymes on Electrodes. In: Guisan J.M., editors. Immobilization of Enzymes and Cells. Methods in Biotechnology™, vol 22. Humana Press. p. 239-50. DOI:10.1007/978-1-59745-053-9_21

[7] Hwang ET, Lee S. Multienzymatic Cascade Reactions via Enzyme Complex by Immobilization. ACS Catalysis. 2019;(9):4402-25. DOI: 10.1021/acscatal.8b04921

[8] Ganesh V, Pandey RR, Malhotra BD, Lakshminarayanan V. Electrochemical characterization of self-assembled monolayers (SAMs) of thiophenol and aminothiophenols on polycrystalline Au: Effects of potential cycling and mixed SAM formation. J Electroanal Chem. 2008;619-620(1-2):87-97. DOI: 10.1016/j.jelechem.2008.03.015

[9] Łuczak T, Osińska M. New self-assembled layers composed with gold nanoparticles, cysteamine and dihydrolipoic acid deposited on bare gold template for highly sensitive and selective simultaneous sensing of dopamine in the presence of interfering ascorbic and uric acids. J Solid State Electrochem. 2017;21(3):747-58. DOI: 10.1007/s10008-016-3416-z

[10] Abad JM, Vélez M, Santamaría C, Guisán JM, Matheus PR, Vázquez L, et al. Immobilization of Peroxidase Glycoprotein on Gold Electrodes Modified with Mixed Epoxy-Boronic Acid Monolayers. J Am Chem Soc. 2002;124(43):12845-53. DOI: 10.1021/ja026658p

[11] Grabowska I, Maes W, Huynh Ngo T, Rohand T, Dehaen W, Radecki J, et al. Multiple redox-active sites in copper dipyrromethene-corrole self-assembled monolayers deposited onto gold electrodes. Int J Electrochem Sci. 2014;9(3):1232-49.

[12] Karimi Shervedani R, Samiei Foroushani M, Bagheri Dehaghi S. Functionalization of gold mercaptopropionic acid self-assembled monolayer with 5-amino-1,10-phenanthroline: Interaction with iron(II) and application for selective recognition of guanine. Electrochim Acta. 2015;164:344-52. DOI: 10.1016/j.electacta.2015.02.170

[13] Mohamad NR, Marzuki NHC, Buang NA, Huyop F, Wahab RA. An overview of technologies for immobilization of enzymes and surface analysis techniques for immobilized

enzymes. Biotechnol Biotechnol Equip. 2015;29(2):205-20. DOI: 10.1080/13102818.2015.1008192

[14] Pinyou P, Blay V, Muresan LM, Noguer T. Enzyme-modified electrodes for biosensors and biofuel cells. Mater Horizons. 2019;1336-58. DOI: 10.1039/c9mh00013e

[15] Carrara S, Baj-Rossi C, Boero C, De Micheli G. Do carbon nanotubes contribute to electrochemical biosensing? Electrochim Acta. 2014;128:102-12. DOI: 10.1016/j.electacta.2013.12.123

[16] Adam C, Scodeller P, Grattieri M, Villalba M, Calvo EJ. Revisiting direct electron transfer in nanostructured carbon laccase oxygen cathodes. Bioelectrochemistry. 2016;109:101-7. DOI: 10.1016/j.bioelechem.2016.01.007

[17] Chen A, Chatterjee S. Nanomaterials based electrochemical sensors for biomedical applications. Chem Soc Rev. 2013;42(12):5425-38. DOI: 10.1039/c3cs35518g

[18] Konrad MP, Doherty AP, Bell SEJ. Stable and Uniform SERS Signals from Self-Assembled Two-Dimensional Interfacial Arrays of Optically Coupled Ag Nanoparticles. Anal Chem. 2013;85(14):6783-9. DOI: 10.1021/ac4008607

[19] Alver E, Metin AÜ. Chitosan based metal-chelated copolymer nanoparticles: Laccase immobilization and phenol degradation studies. Int Biodeterior Biodegradation. 2017;125:235-42. DOI: 10.1016/j.ibiod.2017.07.012

[20] Gao C, Guo Z, Liu J-HH, Huang X-JJ. The new age of carbon nanotubes: An updated review of functionalized carbon nanotubes in electrochemical sensors. Nanoscale. 2012;4(6):1948. DOI: 10.1039/c2nr11757f

[21] Rosca V, Catalin Popescu I. Kinetic analysis of horseradish peroxidase "wiring" in redox polyelectrolyte-peroxidase multilayer assemblies. Electrochem commun. 2002;4(11):904-11. DOI: 10.1016/S1388-2481(02)00486-1

[22] Raghu P, Madhusudana Reddy T, Reddaiah K, Jaidev LR, Narasimha G. A novel electrochemical biosensor based on horseradish peroxidase immobilized on Ag-nanoparticles/poly(l-arginine) modified carbon paste electrode toward the determination of pyrogallol/hydroquinone. Enzyme Microb Technol. 2013;52(6-7):377-85. DOI: 10.1016/j.enzmictec.2013.02.010

[23] Rawal R, Chawla S, Pundir CS. Polyphenol biosensor based on laccase immobilized onto silver nanoparticles/multiwalled carbon nanotube/polyaniline gold electrode. Anal Biochem. 2011;419(2):196-204. DOI: 10.1016/j.ab.2011.07.028

[24] Huerta-Miranda GA, Arrocha-Arcos AA, Miranda-Hernández M. Gold nanoparticles/4-aminothiophenol interfaces for direct electron transfer of horseradish peroxidase: Enzymatic orientation and modulation of sensitivity towards hydrogen peroxide detection. Bioelectrochemistry . 2018;122:77-83. DOI: 10.1016/j.bioelechem.2018.03.004

[25] Smith AT, Veitch NC. Substrate binding and catalysis in heme peroxidases. Curr Opin Chem Biol. 1998;2(2):269-78. DOI: 10.1016/S1367-5931(98)80069-0

[26] Battistuzzi G, Bellei M, Bortolotti CA, Sola M. Redox properties of heme peroxidases. Arch Biochem Biophys. 2010;500(1):21-36. DOI: 10.1016/j.abb.2010.03.002

[27] Fang Y-S, Huang X-J, Wang L-S, Wang J-F. An enhanced sensitive electrochemical immunosensor based on efficient encapsulation of enzyme in

silica matrix for the detection of human immunodeficiency virus p24. Biosens Bioelectron. 2015;64:324-32. DOI: 10.1016/j.bios.2014.09.022

[28] Su X-D, Zhang H, Terwilliger TC, Liljas A, Xiao J, Dong Y. Protein Crystallography from the Perspective of Technology Developments. Crystallogr Rev . 2015;21(1-2):122-53. DOI: 10.1080/0889311X.2014.973868

[29] Helliwell JR. New developments in crystallography: exploring its technology, methods and scope in the molecular biosciences. Biosci Rep . 2017;37(4):BSR20170204. DOI: 10.1042/BSR20170204

[30] Shigematsu H. Electron cryo-microscopy for elucidating the dynamic nature of live-protein complexes. Biochim Biophys acta Gen Subj. 2020;1864(2):129436. DOI: 10.1016/j.bbagen.2019.129436

[31] Bai X, McMullan G, Scheres SHW. How cryo-EM is revolutionizing structural biology. Trends Biochem Sci. 2015;40(1):49-57. DOI: 10.1016/j.tibs.2014.10.005

[32] Amunts A, Brown A, Toots J, Scheres SHW, Ramakrishnan V. Ribosome. The structure of the human mitochondrial ribosome. Science. 2015;348(6230):95-8. DOI: 10.1126/science.aaa1193

[33] Tuszynski JA, Winter P, White D, Tseng CY, Sahu KK, Gentile F, et al. Mathematical and computational modeling in biology at multiple scales Biophysics. Theor Biol Med Model. 2014; 11(1):52. DOI: 10.1186/1742-4682-11-52

[34] Dror RO, Dirks RM, Grossman JP, Xu H, Shaw DE. Biomolecular Simulation: A Computational Microscope for Molecular Biology. Annu Rev Biophys. 2012;41(1):429-52. DOI: 10.1146/annurev-biophys-042910-155245

[35] Lu D, Aksimentiev A, Shih AY, Cruz-Chu E, Freddolino PL, Arkhipov A, et al. The role of molecular modeling in bionanotechnology. Physical Biology. 2006;3(1):S40-S53. DOI: 10.1088/1478-3975/3/1/S05

[36] Ramachandran KI, Deepa G, Namboori K. Computational chemistry and molecular modeling: Principles and applications. 1st ed. Springer; 2008. 397 p. DOI: 10.1007/978-3-540-77304-7

[37] Whitehead TA, Baker D, Fleishman SJ. Computational design of novel protein binders and experimental affinity maturation. In: Methods in Enzymology. Elsevier. 2013;523:1-19.DOI: 10.1016/B978-0-12-394292-0.00001-1

[38] Heinecke A, Eckhardt W, Horsch M, Bungartz HJ. Molecular dynamics simulation. In: SpringerBriefs in Computer Science. Springer. 2015;(523):11-29.

[39] Van Der Kamp MW, Mulholland AJ. Combined quantum mechanics/molecular mechanics (QM/MM) methods in computational enzymology. Biochemistry. 2013; 52(16):2708-2728. DOI: 10.1021/bi400215w

[40] van der Kamp MW, Mulholland AJ. Computational enzymology: insight into biological catalysts from modelling. Nat Prod Rep . 2008;25(6):1001-14. DOI: 10.1039/B600517A

[41] Garcia-Garcia WI, Vidal-Limon A, Arrocha-Arcos AA, Palomares LA, Ramirez OT, Miranda-Hernández M. Rotavirus VP6 protein as a bio-electrochemical scaffold: Molecular dynamics and experimental electrochemistry. Bioelectrochemistry. 2019;127:180-186. DOI: 10.1016/j.bioelechem.2019.02.012

[42] Plascencia-villa G, Mena JA, Castro-acosta RM, Fabián JC, Ramírez OT, Palomares LA. Strategies

for the purification and characterization of protein scaffolds for the production of hybrid nanobiomaterials. 2011;879:1105-11. DOI: 10.1016/j.jchromb.2011.03.027

[43] Carreño-Fuentes L, Plascencia-Villa G, Palomares LA, Moya SE, Ramírez OT. Modulating the physicochemical and structural properties of gold-functionalized protein nanotubes through thiol surface modification. Langmuir. 2014;30(49):14991-8. DOI: 10.1021/la503704a

[44] Humphrey W, Dalke A, Schulten K. VMD: Visual molecular dynamics. J Mol Graph . 1996;14(1):33-8. DOI: 10.1016/0263-7855(96)00018-5

[45] Berglund GI, Carlsson GH, Smith AT, Szöke H, Henriksen A, Hajdu J. The catalytic pathway of horseradish peroxidase at high resolution. Nature. 2002;417(6887):463-8. DOI: 10.1038/417463a

[46] Yang J, Yan R, Roy A, Xu D, Poisson J, Zhang Y. The I-TASSER Suite: protein structure and function prediction. Nat Methods . 2015;12(1):7-8. DOI: 10.1038/nmeth.3213

[47] Olsson MHM, Søndergaard CR, Rostkowski M, Jensen JH. PROPKA3: Consistent Treatment of Internal and Surface Residues in Empirical pKa Predictions. J Chem Theory Comput. 2011;7(2):525-37. DOI: 10.1021/ct100578z

[48] Maier JA, Martinez C, Kasavajhala K, Wickstrom L, Hauser KE, Simmerling C. ff14SB: Improving the Accuracy of Protein Side Chain and Backbone Parameters from ff99SB. J Chem Theory Comput . 2015;11(8):3696-713. DOI:10.1021/acs.jctc.5b00255

[49] Essmann U, Perera L, Berkowitz ML, Darden T, Lee H, Pedersen LG. A smooth particle mesh Ewald method. J Chem Phys. 1995;103(19):8577-93. DOI: 10.1063/1.470117

[50] Case DA, Walker RC, Cheatham TE, Simmerling C, Roitberg A, Merz KM, et al. Amber 18. Univ California, San Fr. 2018.

[51] Miao Y, McCammon JA. Gaussian Accelerated Molecular Dynamics: Theory, Implementation, and Applications. Annu Rep Comput Chem. 2017;13:231-78. DOI: 10.1016/bs.arcc.2017.06.005

[52] Miao Y. Acceleration of biomolecular kinetics in Gaussian accelerated molecular dynamics. J Chem Phys. 2018; 149(7):072308. DOI: 10.1063/1.5024217

[53] Miao Y, McCammon JA. Unconstrained enhanced sampling for free energy calculations of biomolecules: a review. Mol Simul. 2016; 42(13):1046-55. DOI: 10.1080/08927022.2015.1121541

[54] Franco-Ulloa S, Riccardi L, Rimembrana F, Pini M, De Vivo M. NanoModeler: A Webserver for Molecular Simulations and Engineering of Nanoparticles. J Chem Theory Comput. 2019; 15(3):2022-32. DOI: 10.1021/acs.jctc.8b01304

[55] Ozer G, Valeev EF, Quirk S, Hernandez R. Adaptive Steered Molecular Dynamics of the Long-Distance Unfolding of Neuropeptide Y. J Chem Theory Comput. 2010;6(10):3026-38. DOI:10.1021/ct100320g

[56] Bureau HR, Merz DR, Hershkovits E, Quirk S, Hernandez R. Constrained unfolding of a helical peptide: Implicit versus explicit solvents. PLoS One. 2015;10(5):e0127034. DOI: 10.1371/journal.pone.0127034

[57] Miao Y, Feher VA, McCammon JA. Gaussian Accelerated Molecular Dynamics: Unconstrained Enhanced

Sampling and Free Energy Calculation. J Chem Theory Comput. 2015;11(8):3584-95. DOI: 10.1021/acs.jctc.5b00436

[58] Miao Y, Feixas F, Eun C, McCammon JA. Accelerated molecular dynamics simulations of protein folding. J Comput Chem. 2015;36(20):1536-1549. DOI: 10.1002/jcc.23964

[59] Valeur E, Bradley M. Amide bond formation: beyond the myth of coupling reagents. Chem Soc Rev. 2009;38(2):606-31. DOI: 10.1039/B701677H

[60] Montalbetti CAGN, Falque V. Amide bond formation and peptide coupling. Tetrahedron . 2005;61(46):10827-52. DOI: 10.1016/j.tet.2005.08.031

Chapter 7

Energy Minimization

Budhayash Gautam

Abstract

The energetic state of a protein is one of the most important representative parameters of its stability. The energy of a protein can be defined as a function of its atomic coordinates. This energy function consists of several components: 1. Bond energy and angle energy, representative of the covalent bonds, bond angles. 2. Dihedral energy, due to the dihedral angles. 3. A van der Waals term (also called Leonard-Jones potential) to ensure that atoms do not have steric clashes. 4. Electrostatic energy accounting for the Coulomb's Law m protein structure, i.e. the long-range forces between charged and partially charged atoms. All these quantitative terms have been parameterized and are collectively referred to as the 'force-field', for e.g. CHARMM, AMBER, AMBERJOPLS and GROMOS. The goal of energy Minimization is to find a set of coordinates representing the minimum energy conformation for the given structure. Various algorithms have been formulated by varying the use of derivatives. Three common algorithms used for this optimization are steepest descent, conjugate gradient and Newton–Raphson. Although energy Minimization is a tool to achieve the nearest local minima, it is also an indispensable tool in correcting structural anomalies, viz. bad stereo-chemistry and short contacts. An efficient optimization protocol could be devised from these methods in conjunction with a larger space exploration algorithm, e.g. molecular dynamics.

Keywords: energy minimization, minimum energy conformation, force fields, global minimum energy, molecular dynamics simulations, molecular modeling

1. Introduction

Molecular modeling relies on the event of theoretical and computational methodologies, to model and study the behavior of molecules, from little chemical systems to big biological molecules and material assemblies. The applying fields of molecular modeling regard computational chemistry, drug design, computational biology and materials science. The fundamental computational technique to perform molecular modeling is simulation. Molecular simulation techniques need specific extra computational and code software system [1]. Most molecular modeling studies involve three stages. Within the initial stage a model is chosen to explain the intra- and inter-molecular interaction within the system. The two most typical models that are utilized in molecular modeling are quantum mechanics and molecular mechanics. These models enable the energy of any arrangement of the atoms and molecules within the system to be calculated and permit the modeler to work out how the energy of the system varies because the positions of the atoms and molecules change. The second stage of a molecular modeling study is that the calculation itself, such as an energy Minimization, a molecular dynamics or Monte Carlo simulation, or a Conformations search. Finally, the calculation should be analyzed, not solely to calculate properties however additionally to see that it's been performed properly [2].

In molecular modeling we tend to are particularly curious about minimum points on the energy surface. Minimum energy arrangements of the atoms correspond to stable states of the system; any movement off from a minimum provides a configuration with a better energy. There is also a really sizable amount of minima on the energy surface. The minimum with the very lowest energy is known as the global energy minimum. To spot those geometries of the system that correspond to minimum points on the energy surface we tend to use a Minimization algorithm. The highest point on the pathway between two minima is of particular interest and is understood as the saddle point, with the arrangement of the atoms being the transition structure. Both minima and saddle points are stationary points on the energy surface, wherever the primary derivative of the energy function is zero with regard to all the coordinates [3].

A geographical analogy is useful thanks to illustrate several of the ideas as during this analogy minimum points correspond to the lowest of valleys. A minimum is also represented as being in an exceedingly 'long and slender valley' or 'a flat and featureless plain'. Saddle points correspond to mountain passes. Confer with consult with algorithms creating steps as 'uphill' and downhill'.

1.1 Energy minimization: a brief description about the problem

The Minimization problem can be formally stated as follows: given a function *f* which depends on one or more independent variables $x_1, x_2, \ldots, x_i$, find the values of those variables where *f* has a minimum value. At a minimum point the first derivative of the function with respect to each of variables is zero and the second derivative are all positive:

$$\partial f / \partial x_i = 0; \quad \partial^2 f / \partial x_i^2 > 0 \qquad (1)$$

With respect to present discussion, the most important functions are the quantum mechanics or molecular mechanics energy with the variables x_i being the Cartesian or the internal co-ordinates of the atoms. It is a common practice to always perform Molecular mechanics Minimizations in Cartesian co-ordinates, in which the energy is a function of *3 N* variables; on the other hand, for quantum mechanics internal co-ordinates are often used. The least value of any function can be identified using standard calculus methods for analytical functions. But, due to the complexities of pattern of energy change with change in the coordinates, it is almost impossible for any molecular system. Therefore, the energy minima are often identified with the help of numerical methods. These methods gradually make changes to the coordinates to generate configurations having lower and lower energies until the minimum is reached [2].

Minimization algorithms can be classified into two categories: one in which we use derivatives of the energy with respect to the coordinates and second in which we do not use any derivative. Derivatives are extremely important because they have details about the shape of the energy surface and due to this efficiency to locate the minimum energy is increases drastically. For any problem, before choosing best algorithm (or algorithms), several points should be considered for e.g. the best Minimization algorithm should use least memory to generate the answer as quickly as possible. For different problems of molecular modeling, different Minimization method is used because we do not have any Minimization method developed which could be applied to all. Any method which is developed for efficient performance with quantum mechanics may or may not be compatible for molecular mechanics because quantum mechanics deals with models having very less atoms as compare to molecular mechanics. Another point is that procedures like inversion of matrix in some Minimization methods works fine for small systems but problem arises when number of atoms increases to thousands. To calculate the number of derivatives of different Conformations and their energies, different level of performance is required for quantum mechanics than molecular mechanics. Molecular mechanics requires an algorithm that is having more number of steps; quantum mechanics has the opposite scenario. Therefore we have various methods in various popular software packages [4].

As, most of the Minimization algorithms can only identify the minimum energy point which is closest to the starting point, thus it can be stated that they only move downwards or more appropriately downhill on the energy surface. Suppose, this schematic energy surface is shown in **Figure 1**, having three starting points A, B and C to obtained the minima. The locations at which any hypothetical ball stops rolling on energy surface under gravitational force will have corresponding energy minima. But more important thing is to identify global energy minimum which can only be generated by using different starting points, which will be minimized later. Using this criterion, some of the Minimization methods can move uphill to find out energy minimum than the closest one. But not a single algorithm till date is reported for efficiently identification of the global minimum energy from a random starting point. To identify the number of different minimum energy Conformations the shape of energy surface is very useful. For example, population or number of minimum in a deep and narrow valley will be very less than population at broad minimum because it is having higher energy as the vibrational energy is more widely spaced in the minimum and so less accessible. Therefore, the global energy minimum may not be the most highly populated minimum. Thus, there may be the case that the 'functional' structure (e.g. the biologically active conformation of a drug molecule) may not belong to the global minimum, or to the most highly populated conformation, or even to a minimum energy structure at all [3].

Every Minimization algorithm has a set of initial coordinates as input. These coordinates can be generated from different sources. These can be generated using traditional experimental method like X-ray crystallography or NMR. Alternatively, these can be generated by employing a theoretical method like conformational search procedure. But for practical efficiency, both types of methods can be used combinatorially. For example, the behavior of a protein in water can be studied

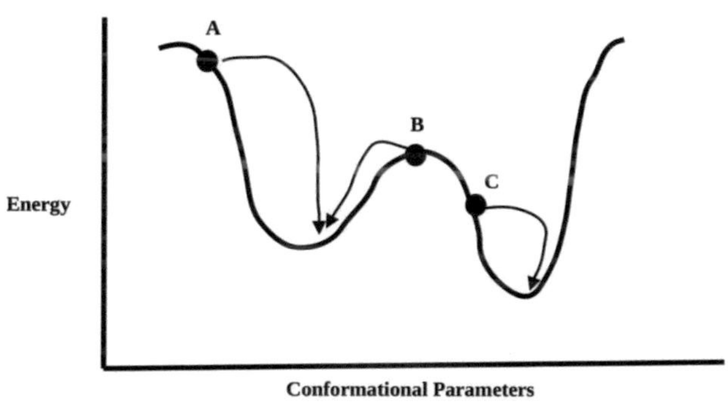

Figure 1.
A one-dimensional energy surface showing minimization methods movement downwards or downhill towards the closest energy minimum.

using its x-ray generated structure. Then place this in a solvent completely. Monte Carlo or molecular dynamics simulation can generate the atomics or Cartesian coordinates of the solvent molecules.

1.2 Derivatives

For derivatives based Minimization methods, calculation of the derivatives of the energy is performed with respect to the different variables i.e. Cartesian or internal coordinates, as the case may be. These derivatives can be generated using either analytical or numerical procedures but derivatives obtained through analytical procedure are more preferred because these can be generated more readily and these are more exact. Although, if derivatives generated by only numerical procedure is available then one should use a non-derivative Minimization procedure as it is more efficient [3].

Although, in some situations it is always preferable to use derivatives generated though numerical procedure. By following way these can be generated: suppose there is a small alteration (δx_i) in one of the coordinates x_i and the energy calculation is performed for this new alteration the by dividing the alteration in energy (δE) by the alteration in coordinate ($\delta E / \delta x_j$), the derivative $\partial E/\partial x_i$ is obtained. This rigorously yields the derivative at the mid-point between the two points x_i and $x_i + \delta x_i$. A more correct value of the derivative at the point x_i; could also be acquired (at the price of a further energy calculation) by assessing the energy at two points, $x_i + \delta x_i$ and $x_i - \delta x_i$. The derivative is then obtained by dividing the variation with in the energies by $2\delta x_i$.

2. Non-derivative minimization methods

2.1 The simplex method

A geometrical figure having $M + 1$ interconnected vertices is called as *simplex*, where dimensionality of the energy function is M. Thus, a simplex with function of two variables will have a triangular shape. Further, for a function of three variables simplex will have tetrahedral shape. Therefore, for an energy function of $3N$ Cartesian coordinates the simplex will have $3N + 1$ vertices; but simplex will have $3N - 5$ vertices, if internal coordinates are used. The energy could be calculated for a specific set of coordinates correspond to each every vertex. For the function $f(x,y) = x^2 + 2y^2$ the simplex method would use a triangular simplex [5].

The simplex algorithm identifies an energy minimum by traveling around on the potential energy surface in a manner that is similar to the movement of an amoeba. There are three possible primary moves. The most common move is a reflection of the vertex having maximum value on the opposite sides of the simplex. The reflection is used as an effort to produce a new point having a lower value. If this is the lowest energy point than any other points in the simplex then next move may be applied which is a "reflection and expansion." Reflection move will be failed to generate a better point, when a "floor of the valley" is reached. In this situation, simplex will make simple contraction all along the highest point dimension. If this fails to further decrease the energy then another kind of move is possible. In this move, contractions occur in all the directions towards the lowest point. The **Figure 2** illustrates above discussed three moves.

The vertices of the initial simplex have to be first generated before applying the simplex algorithm. The first conformation of the method fit to just one of these vertices. Rest of the points can be generated using various methodologies, e.g.

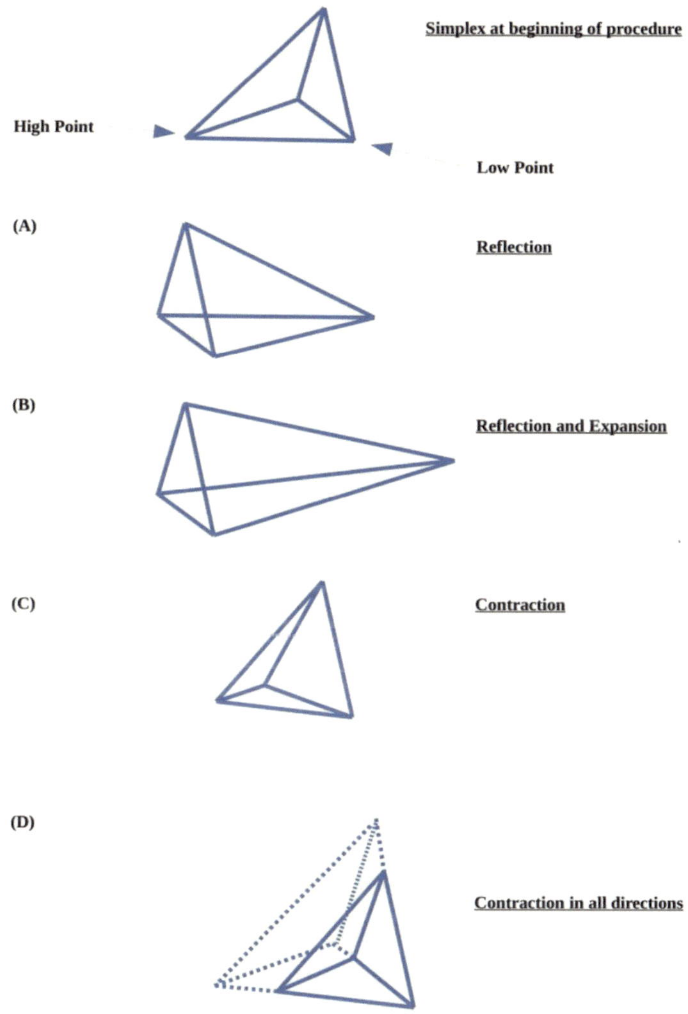

Figure 2.
The three basic moves permitted to the simplex algorithm (reflection, and its close relation reflect-and-expand; contract in one dimension and contract around the lowest point).

simplest method is to increase a fix value to each coordinate successively. To calculate the functional value of the applicable vertex, the energy of the whole system is measured for each new point.

When the starting configuration of the system is having high energy, it is best to use simplex method. The simplex method is more helpful in this because it seldom go wrong in the identification of a fitter answer. Nonetheless, it requires large computational time for the analysis of the high number of energy instances. For e.g. to create the starting simplex needs $3N + 1$ energy analysis. Due to this, the simplex method is frequently used along with other Minimization algorithms. In practice, starting configuration is fine tuned with few steps of the simplex method and then a more suitable and efficient method can be used for further calculations [6].

An important question is that what is the reason behind containing one extra vertex in the simplex than the degree of freedom? The answer to this is that of simplex is having lesser vertices than $M + 1$ then the simplex algorithm cannot search the entire surface of the energy. For e.g. if the simplex having just two vertex (a simplex with only two vertices is simply a straight line) is being used to search the quadratic surface of the energy, the only available move in this scenario would be

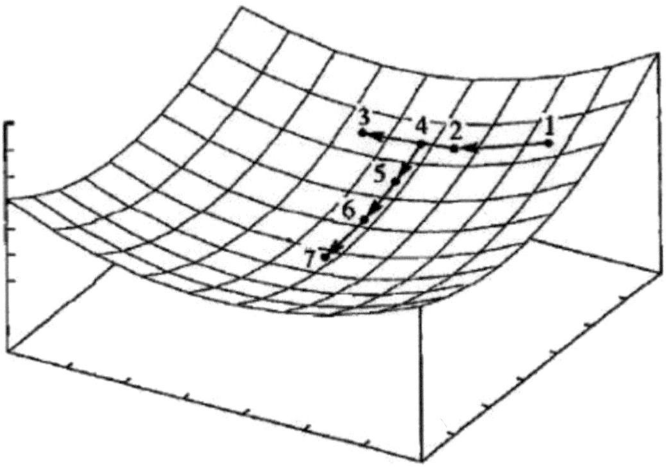

Figure 3.
The sequential univariate search procedure. From the starting point 1, two steps are created along one of the coordinates to give points 2 and 3. A parabola is fitted to these three points and the minimum located (point 4). The same steps is then repeated along the next coordinate (points 5, 6 and 7) (Figure adapted from Schlegel H B 1987. Optimization of equilibrium geometries and transition structures In Lawley K P (editor) ab initio methods in quantum chemistry - I New York, John Wiley, pp. 249–286).

to find out other points that lie on this straight line. In this case, the energy surface which is away from the straight line would not be searched. Likewise, if we have function of three variable and simplex is just a triangle then only the area of search space that lies in the same plane as to the triangle will only be searched, whereas the energy minimum may not be present at this plane [7].

2.2 The sequential univariate search method

It is seldom appropriate to use the simplex method for the calculations involve in quantum mechanics because in that case very high number of energy assessments has to be done. In this case a much befitting non-derivative procedure like the sequential univariate search method is well-advised [8]. This procedure consistently repeats through the coordinates successively. For every coordinate, two new configurations are created by making changes in the present coordinates (i.e. $x_j + \delta x_i$, and $x_i + 2\delta x_i$). Then the energy calculation for these two configurations is performed. Three points related to the two twisted configurations and the original one are then fitted with a parabola. The identification of the minimum point in the current quadratic function is performed. Then in the next step, the coordinate is twisted to the point of the minimum. The procedure is illustrated in **Figure 3**.

The minimum is bound to reach when the changes in all the coordinates are adequately very small. Alternatively, a new iteration is performed. In comparison to the simplex method, the sequential univariate method normally needs less function assessment. But if two or more coordinates have a strong connection or bonding then the sequential univariate search method may converge slowly. It also converges slowly when the energy surface is similar to a long narrow valley.

3. Derivative minimization methods

Most of the favorite Minimization procedures utilize derivatives because the information which is helpful in minimization is furnished by derivatives. The direction of the first derivative of the energy (the gradient) points where the minimum lies and the magnitude of the gradient tells about the steepness of the local slope. The energy of the system can be decreased by moving each atom with respect to the force acting on it. The force is equal to minus the gradient. Second derivatives point towards the curvature of the function. This information can be utilized to find out where the function will change its direction (i.e. pass through a minimum or any different non-moving point).

The energy functions which are often utilized in molecular modeling are seldom quadratic and thus the Taylor series expansion can only be a well-advised approximation. There are two crucial consequences of this. First is, for a pure quadratic function a given minimization procedure executes very well rather than for a molecular mechanics or quantum mechanics energy surface. For example, the Newton–Raphson algorithm can identify the minimum in a one step for a purely quadratic function. But, for a typical molecular modeling energy function, it needs to run several iterations. The second consequence is that, even though they may function very well close to a minimum, where the harmonic approximation is more logical, the harmonic approximation is very bad and far from minimum. Due to this some of the less robust methods will not be successful. Because of this reason Minimization protocol must be picked very carefully. A robust or may be inefficient method could be exploited earlier then a comparatively least robust but more efficient procedure.

On the basis of highest order derivatives used, the derivative methods can be classified. The first derivatives or the gradients based methods are called as first-order methods. Methods in which both first and second order derivatives are used are known as second-order methods. Because the simplex method does not use any derivatives can thus be called as a zeroth-order method.

3.1 First-order minimization methods

The *steepest descents* and the *conjugate gradient* method are two such first order Minimization algorithms which are very often used in molecular modeling. In these methods coordinates of the atoms are altered step by step with respect to their movement towards the minimum point. For each iteration (k), the initial point is the molecular conformation generated from last step. It is represented by the multidimensional vector xk - 1. For the first iteration, the starting point is the initial configuration of the system provided by the user, the vector $x1$.

3.1.1 The steepest descents method

The steepest descents method moves in the direction parallel to the net force, which in our geographical analogy corresponds to walking straight downhill. For $3\ N$ Cartesian coordinates this direction is most conveniently represented by a $3\ N$-dimensional unit vector, sk. Thus:

$$s_k = -g_k / |g_k| \tag{2}$$

Once the direction of movement is clearly characterized then it should be decided that how much distance to be covered along the gradient. Consider the two-dimensional energy surface of **Figure 4.** The gradient direction from the initial point is along the line shown. Suppose we have a cross-section through the surface along the line, the function will pass through a minimum and then increase, as shown in the figure [9]. We can identify the minimum point by performing a line search or we can take a step of arbitrary size along the direction of the force.

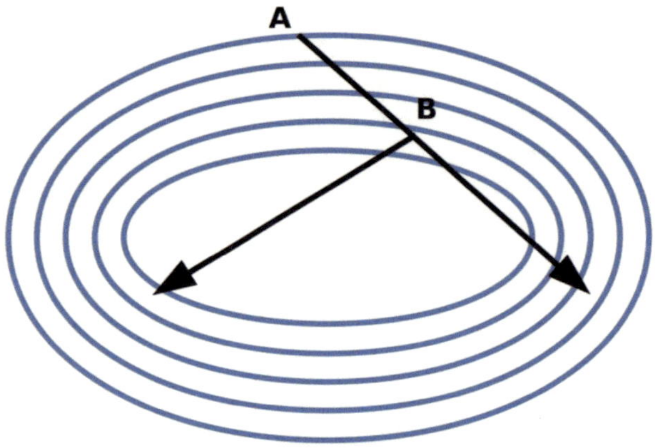

Figure 4.
Steepest descents method.

3.1.2 Line search in one dimension

The goal of a line search is to find out the minimum along a specific direction (i.e. along a line through the multidimensional space) [10]. In the very first step of the line search is to *bracket* the minimum. This implies determining three points along the line in a way that the energy of the intermediate point is less than the energy of the two extrinsic points. If it is possible to identify these kinds of three points, then it should be make sure that two extrinsic points must have at least one minimum in between. Then to reduce the distance in between the three points, an iterative algorithm could be applied which in a step by step manner, limits the minimum to a very smaller space. Theoretically, it looks easy but it may involve a large number of functional analysis. Thus it is computationally very expensive methods.

Alternatively, we can set a suitable quadratic function to the three points. Then apply differentiation to this suited function to modify an approximation to the minimum along the line which should be identified analytically. To get a better approximate, a new function can be set then, as shown in **Figure 5**. Higher-order polynomials may yield an improved fit to the bracketing points but when these

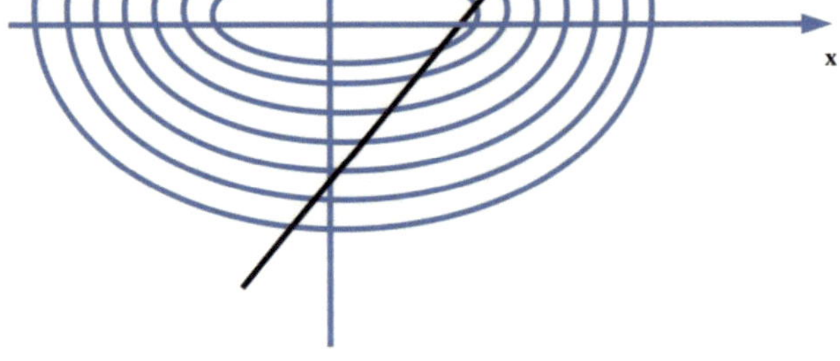

Figure 5.
A line search is used to locate the minimum in the function in the direction of the gradient.

are utilized with functions that altered aggressively in the bracketed region, these higher-order polynomials can yield wrong interpolations. The gradient at the minimum point obtained from the line search will be perpendicular to the previous direction. Thus, when the line search method is used to locate the minimum along the gradient then the next direction in the steepest descents algorithm will be orthogonal to the previous direction (i.e. $gk.\ gk-1=0$) [11].

3.1.3 Arbitrary step approach

As, we know that the line search may be computationally very expensive, New coordinates can be identified by walking a step of arbitrary size along the gradient unit vector sk. The new set of coordinates after step k would then be given by the equation:

$$xk_{+1} = x_k + \lambda_k s_k \qquad (3)$$

where, λ_k is the step size. In most of the applications within molecular modeling, the steepest descents algorithm, the step size at the start has a predetermined default value. If energy decreases after the first iteration, then for second iteration the step size is increases by an increasing component. The process repeats till the point at which each iteration decreases the energy. When a step produces an addition in energy, it is assumed that the algorithm has leapt across the valley which comprise the minimum and up the slope on the opposite face. The step size is then reduced by a multiplicative factor (e.g. 0.5). Often, the size of the step is decided according to the nature of the energy surface. It would be more suitable to have bigger step size for a plane or flat surface rather than a slender or narrow altered valley, where more smaller step are much appropriate. Computational time is less in the case of the arbitrary step method than much stringent line search method, because the arbitrary or random step approach needs higher number of steps to find out the minimum than line search method but arbitrary step method may frequently needs lesser functional analysis [12].

The largest inter-atomic forces indicate the direction of the gradient. Therefore, the steepest descent is more suitable for alleviating attributes of the highest-energy in the initial conformation. If the harmonic calculations corresponding to the energy is hypothesized badly and the initial point is distant from a minimum, even then the method performs strongly. But, in the case of downward movements in a long slender valley, the method uses short steps in high number and this causes trouble to the method. Although, it is not suitable manner to find out the minimum, the steepest descents process is bound to move in the right-angled direction at every point. The route constantly over compensates itself and vibrates. However, Subsequent steps reintroduce errors which were already rectified by prior steps [13].

3.1.4 Conjugate gradients minimization

The vibrating activity of the steepest descents procedure in slender depression is absent in the set of directions generated by the conjugate gradients methods. Rather, both the directions of consecutive steps and the gradients are orthogonal in the steepest descents method [14]. More specifically, in the conjugate gradients method, the gradients are orthogonal in nature at every point and the directions of consecutive steps are *conjugate* that is why it is more correctly known as the conjugate direction method. Because of the feature of a set of conjugate directions, for a quadratic function of M variables, in M steps the minimum can be identified. The conjugate gradients method moves in a direction v_k from point x_k where v_k is computed from the gradient at the point and the previous direction vector v_{k-1}.

Both the conjugate gradients method and the steepest descents method move in the direction of the gradient in the first step. The line search method should ideally be used to find out the one-dimensional minimum in all direction to assure that each gradient is orthogonal to all preceding gradients and that each direction is conjugate to all preceding directions. However, at this stage random step procedure is also achievable [15]. To identify the second point a line search should be applied along the line with gradient but it must pass through the point. Therefore, the conjugate gradient procedure identifies the perfect minimum of the function in just two moves.

3.2 Second order derivative methods

3.2.1 The Newton-Raphson method

Second-order methods utilize the information from both the first derivatives and the second derivatives to find out a minimum. First derivatives provide gradient information while second derivative furnish details about the curvature of the function. Having these properties, the Newton–Raphson method is the simplest second order method [16]. For a strictly quadratic function of the first derivative the second derivative will be same everywhere. If we talk about a multidimensional function the Hessian matrix of second derivatives essentially be inverted. Thus, for larger molecules it is more computationally expensive because there are a large number of atoms present and this necessitates bigger storage. The Newton-Raphson method is thus more appropriate to small molecules (usually less than 100 atoms or so) [17].

As stated earlier, for a strictly quadratic function, the Newton–Raphson method requires just one step to locate the minimum from any point on the surface. Practically, the surface is exclusively quadratic to a first approximation and this necessitates a large number of steps to move. The Hessian matrix of second derivatives should be calculated first and then inverted at each step. This must be 'positive definite' in a Newton–Raphson Minimization method. A positive definite matrix is one for which all the eigen-values are positive. When the Hessian matrix is not positive definite then the Newton–Raphson method moves to saddle points where the energy increases, rather than a narrow point where energy decreases. Additionally, the harmonic approximation is not suitable at positions which are very far from the minimum because this leads to instability of the Minimization. This can be solved by employing a more efficient and robust method (prior to the application of the Newton–Raphson method) to find out minimum or to reach close to minimum (in case of the positive definite Hessian matrix) [18].

3.2.2 Quasi-Newton method

Computation of the inverse Hessian matrix can be a possibly long procedure that represents an important disadvantage to the 'pure' second derivative methods such as Newton–Raphson. Furthermore, analytical second derivatives could not be generated preferably. Variable metric methods which are also an alternative name to the Quasi-Newton methods gradually develop the inverse Hessian matrix in consecutive iterations. That means, a sequence of matrices H_k is developed.

At each iteration k, the new positions x_{k+1} are obtained from the current positions x_k, the gradient g_k and the current approximation to the inverse Hessian matrix H_k. For quadratic function it is same, but for 'real' job a line search may be desired. Hence, a line search is performed along the vector $(x_{k+1} - x_k)$. It may not be essential to find out the minimum in the direction of the line search very accurately, at the cost of a few more steps of the quasi-Newton algorithm [19]. For quantum mechanics calculations the additional energy evaluations required by the line search may prove more expensive than using the more approximate approach. An effective compromise is to fit a function to the energy and gradient at the current point x_k and at the point x_{k+1} and find out the minimum in the fitted function [20].

4. Which minimization method should 1 use?

The choice of Minimization algorithm is determined by a number of components, including the storage and computational requirements, the relative speeds with which the various parts of the calculation can be performed, the availability of analytical derivatives and the robustness of the method. Thus, any method that requires the Hessian matrix to be stored (let alone its inverse calculated) may present memory problems when applied to systems containing thousands of atoms. Calculations on systems of this size are invariably performed using molecular mechanics, and so the steepest descents and the conjugate gradients methods are very popular here. For molecular mechanics calculations on small molecules, the Newton–Raphson method may be used, although this algorithm can have problems with structures that are far from a minimum. For this reason it is usual to perform a few steps of Minimization using a more robust method such as the simplex or steepest descents before applying the Newton–Raphson algorithm Analytical expressions for both first and second derivatives are available for most of the terms found in common force fields. The steepest descent method can actually be superior to conjugate gradients when the starting structure is some way from the minimum. However, conjugate gradients are much better once the initial strain has been removed. Quantum mechanical calculations are restricted to systems with relatively small numbers of atoms, and so storing the Hessian matrix is not a problem. As the energy calculation is often the most time-consuming part of the calculation, it is desirable that the Minimization method chosen takes as few steps as possible to reach the minimum. For many levels of quantum mechanics theory analytical first derivatives are available. However, analytical second derivatives are only available for a few levels of theory and can be expensive to compute. The quasi-Newton methods are thus particularly popular for quantum mechanical calculations.

When using internal coordinates in a quantum mechanical Minimization it can be important to use an appropriate Z-matrix as input. For many systems the Z-matrix can often be written in many different ways as there are many combinations of internal coordinates. There should be no strong coupling between the coordinates. Dummy atoms can often help in the construction of an appropriate Z-matrix. A dummy atom is used solely to define the geometry and has no nuclear charge and no basis functions. Strong coupling between coordinates can give long 'valleys' in the energy surface, which may also present problems. Care must be taken when defining the Z-matrix for cyclic systems in particular. The natural way to define a cyclic compound would be to number the atoms sequentially around the ring. However, this would then mean that the ring closure bond will be very strongly coupled to all of the other bonds, angles and torsion angles. Some quantum mechanics programs are able to convert the input coordinates (be they Cartesian or internal) into the most efficient set for Minimization so removing from the user the problems of trying to decide what is an appropriate set of internal coordinates. For energy Minimizations redundant internal coordinates have been shown to give significant improvements in efficiency compared with Cartesian coordinates or non-redundant internal coordinates, especially for flexible and polycyclic systems [21]. The redundant internal coordinates employed generally com- comprise the bond lengths, angles and torsion angles in the system. These methods obviously also require the means to inter-convert between the internal coordinate representation and the Cartesian coordinates that are often used as input and desired as output. Of particular importance is the need to transform energy derivatives and the Hessian matrices (if appropriate) [22].

5. Differentiating between minima, maxima and saddle points

A configuration at which all the first derivatives are zero need not necessarily be a minimum point; this condition holds at both maxima and saddle points as well. From simple calculus we know that the second derivative of a function of one variable, f'(x) is positive at a mini- minimum and negative at a maximum. It is necessary to calculate the eigenvalues of the Hessian matrix to distinguish between minima, maxima and saddle points. At a minimum point there will be six zero and 3 N — 6 positive eigenvalues if 3 N Cartesian coordinates are used. The six zero eigenvalues correspond to the translational and rotational degrees of free- freedom of the molecule (thus these six zero eigenvalues are not obtained when internal coordinates are used). At a maximum point all eigenvalues are negative and at a saddle point one or more eigenvalues are negative.

6. What should be the convergence criteria?

In contrast to the simple analytical functions that we have used to illustrate the operation of the various Minimization methods, in 'real' molecular modeling applications it is rarely possible to identify the 'exact' location of minima and saddle points. We can only ever hope to find an approximation to the true minimum or saddle point. Unless instructed otherwise, most Minimization methods would keep going forever, moving ever closer to the minimum. It is therefore necessary to have some means to decide when the Minimization calculation is sufficiently close to the minimum and so can be terminated. Any calculation is of course limited by the precision with which numbers can be stored on the computer, but in most instances it is usual to stop well before this limit is reached. A simple strategy is to monitor the energy from one iteration to the next and to stop when the difference in energy between successive steps falls below a specified threshold. An alternative is to monitor the change in coordinates and to stop when the difference between successive configurations is sufficiently small. A third method is to calculate the root-mean-square gradient. This is obtained by adding the squares of the gradients of the energy with respect to the coordinates, dividing by the number of coordinates and taking the square root. It is also useful to monitor the maximum value of the gradient to ensure that the Minimization has properly relaxed all the degrees of freedom and has not left a large amount of strain in one or two coordinates [23].

7. Applications of energy minimization

Energy Minimization is very widely used in molecular modeling and is an integral part of techniques such as conformational search procedures. Energy Minimization is also used to prepare a system for other types of calculation. For example, energy mini- Minimization may be used prior to a molecular dynamics or Monte Carlo simulation in order to relieve any unfavorable interactions in the initial configuration of the system [24]. This is especially recommended for simulations of complex systems such as macromolecules or large molecular assemblies.

8. Conclusion

The energetic state of a protein is one of the most important representative parameter of its stability. The energy of a protein (E) can be defined as a function of its atomic coordinates, thus providing a quantitative criterion for model selection and refinement. This energy function consists of several components e.g. (1) Bond energy and angle energy, representative of the covalent bonds, bond angles. (2) Dihedral energy, due to the dihedral angles. (3) A van der Waals term (also called Leonard-Jones potential) to ensure that atoms do not have steric clashes. (4) Electrostatic energy accounting for the Coulomb's Law m protein structure, i.e. the long-range forces between charged and partially charged atoms. All these quantitative terms have been parameterized and are collectively referred to as the 'forcefield'. The goal of energy Minimization is to find a set of coordinates representing the minimum energy conformation for the given structure. Various algorithms have been formulated by varying the use of derivatives. The common algorithm used for this optimization is steepest descent, conjugate gradient and Newton–Raphson etc. These methods complement each other in search of the local minima. Therefore, a reasonable energy Minimization protocol involves few initial steps of steepest descent, followed by a larger number of conjugate gradient iterations. Although energy Minimization is a tool to achieve the nearest local minima, it is also an indispensable tool in correcting structural anomalies, viz. bad stereo-chemistry and short contacts. An efficient optimization protocol could be devised from these methods in conjunction with a larger space exploration algorithm, e.g. molecular dynamics.

Acknowledgements

The authors are grateful to the Sam Higginbottom University of Agriculture, Technology and Sciences, Allahabad, for providing the facilities and support to complete the present work.

Conflict of interest

The author declares no conflict of interest.

Author details

Budhayash Gautam
Sam Higginbottom University of Agriculture, Technology and Sciences, Prayagraj, U. P., India

*Address all correspondence to: budhayashgautam@gmail.com

References

[1] Ebejer Jean-Paul, Fulle Simone, Morris Garrett M., Finn Paul W. The emerging role of cloud computing in molecular modelling, Journal of Molecular Graphics and Modelling, 2013; 44,177-187.

[2] *Adcock* S A, *McCammon* J A. *Molecular dynamics*: *survey* of *methods* for *simulating* the *activity* of *proteins*. *Chem Rev*. *2006;106:1589-1615*.

[3] Hinchliffe Alan. Molecular Modelling for Beginners, 2nd Edition. John Wiley & Sons Ltd. 2003.

[4] Leach AR. Molecular Modelling: Principles and Applications. Prentice Hall, 2001.

[5] Thomas H. Cormen, Charles E. Leiserson, Ronald L. Rivest, Clifford Stein. Introduction to Algorithms, Second Edition. MIT Press and McGraw-Hill, 2001;790-804.

[6] Disser, Yann, Skutella Martin. The Simplex Algorithm Is NP-Mighty. ACM Trans. Algorithms. 2018;15(1):5:1-5:19.

[7] *Maros István, Mitra Gautam. Simplex algorithms. In J. E. Beasley (ed.). Advances in linear and integer programming. Oxford Science. 1996;1-46.*

[8] Soumitra Sitole. Univariate Search Method (Optimizing Quadratic Equations with Two Variables), MATLAB Central File Exchange. 2020.

[9] Wiberg K J. A Scheme for Strain Energy Minimization. Application to the Cycloalkanes. Am. Chem. Soc., 1965; 87,5,1070-78.

[10] Nocedal Jorge, Wright Stephen J. Line Search Methods. Numerical Optimization. New York: Springer. 1999; 34-63.

[11] Wenyu Sun, Ya-Xiang Yuan. Line Search. Optimization Theory and Methods: Nonlinear Programming. New York: Springer. 2006;71-117.

[12] Elber R, Meller J , Olender R. Stochastic Path Approach to Compute Atomically Detailed Trajectories: Application to the Folding of C Peptide *The Journal of Physical Chemistry B*. **1999;***103*(6),899-911.

[13] Hsieh W, Kuo M, Yau H F, Chang Chi Ching. A simple arbitrary phase-step digital holographic reconstruction approach without blurring using two holograms. OPT REV. 2009;16,466-471.

[14] Fletcher R, Reeves C M. Function Minimization by conjugate gradients. *The Computer Journal. 1964;*7,2,149-154.

[15] *Polak E, Ribiere G. Rev. Fr.* Infolrm. *Rech. Operations 16R1,1969;35*-43.

[16] Ypma, Tjalling J. Historical development of the Newton–Raphson method, SIAM Review. 1995;37(4), 531-551.

[17] Gil A, Segura J, Temme N M. Numerical methods for special functions. Society for Industrial and Applied Mathematics (SIAM). 2007; http://dx.doi.org/10.1137/1.9780898717822

[18] *Süli Endre, Mayers David. An Introduction to Numerical Analysis. Cambridge University Press, 2003.*

[19] *Broyden C G. Quasi-Newton Methods. In Murray, W. (ed.). Numerical Methods for Unconstrained Optimization. London: Academic Press. 1972;pp.87-106.*

[20] Haelterman Rob, Eester Dirk Van, Verleyen Daan. Accelerating the solution of a physics model inside a tokamak using the (Inverse) Column Updating Method. Journal of Computational and Applied Mathematics. 2015;279:133-144.

[21] Peng C, Ayala PY, Schlegel H B, Frisch M J. Using Redundant Internal Coordinates to Optimise Equilibrium Geometries and Transition States. Journal of Computational Chemistry. 1996;17:49-56.

[22] Westheimer F H. Steric Effects in Organic Chemistry, ed. M. S. Newman, John Wiley & Sons, New York, 1956.

[23] Hendrickson J B. Molecular Geometry. I. Machine Computation of the Common Rings. J. Am. Chem. Soc.1961; 83, 22, 4537-47.

[24] Schlegel H B. Optimization of Equilibrium Geometries and Transition Structures In Lawley K P (Editor) Ab Initio Methods in Quantum Chemistry - I New York, John Wiley, 1987; 249-286.